RIVERSIDE COMMUNITY COLLEGE
1916

STATISTICAL METHODS IN ENGINEERING AND MANUFACTURING

STATISTICAL METHODS IN ENGINEERING AND MANUFACTURING

JOHN E. BROWN

Quality Press
Milwaukee

STATISTICAL METHODS IN ENGINEERING AND MANUFACTURING

JOHN E. BROWN

Library of Congress Cataloging-in-Publication Data

Brown, John E.
Statistical methods in engineering and manufacturing / John E. Brown.
p. cm.
Includes bibliographical references.
ISBN 0-87389-077-9
1. Engineering--Statistical methods. 2. Manufacturing processes--Statistical methods. I. Title.
TA340.B76 1990
001.4'22--dc20 89-18041
CIP

10987654321

ISBN 0-87389-077-9

Acquisitions Editor: Jeanine L. Lau
Production Editor: Tammy Griffin
Cover design by Artistic License. Set in Franklin Gothic by DanTon Typographers.
Printed and bound by Edwards Brothers.

Printed in the United States of America

Quality Press, American Society for Quality Control
310 West Wisconsin Avenue, Milwaukee, Wisconsin

CONTENTS

Preface . . . ix

Abstract . . . xi

SECTION 1 — Frequency Distributions and Probability Curves

- Descriptive and Inferential Statistics . . . 3
- Defining Frequency Distributions . . . 4
- Probability Curves . . . 4
- Binomial Distribution . . . 6
- Poisson Distribution . . . 7
- t Distribution . . . 8
- Chi Square (χ^2) Distribution . . . 9
- F Distribution . . . 9

SECTION 2 — Attribute Sampling

- Attribute Sampling Definition . . . 13
- Binomial Sampling Probabilities . . . 13
- Binomial Sampling Plans . . . 14
- Operating Characteristic Curves . . . 15
- Poisson Sampling Plans . . . 18
- Sampling Risks . . . 19

SECTION 3 — Normal Distributions

- The Normal Curve . . . 29
- Measures of Central Tendency . . . 29
- Measures of Dispersion . . . 30
- Moments About the Mean . . . 30
- Calculating Standard Deviation . . . 31
- Area Under the Normal Curve . . . 32
- The z Score . . . 34

SECTION 4 — Statistical Control Charts

- Detection versus Prevention . . . 43
- Process Capability . . . 43
- Process Stability . . . 45
- Subset Sampling . . . 45
- Control Chart for Averages . . . 48
- Control Chart for Range . . . 50

Calculating Control Limits 50
Control Chart Interpretation 53
Control Charts for Individual Measurements 57
Control Charts for Attributes 59
Control Chart for Percent Defective 60
Control Chart for Number Defective: np Chart 63
Control Chart for Defects: c Chart 63
Control Chart for Average Defects: u Chart 65
Control Chart for Demerits: D Chart 67

SECTION 5 — Data Gathering, Presentation, and Basic Problem Solving

Purpose of Data 81
Frequency Distributions 81
Establishing Class Intervals 82
Class Interval Limits and Midpoints 84
Histograms 85
Frequency Polygon 85
Cumulative Frequency Polygons 89
Trend Charts 90
Bar Charts 93
Scatter Charts 95
Sequential Frequency Chart 99
Graphical Aids in Problem Solving 101
  Pareto Chart 101
  Cause-Effect Diagrams 104
  Scatter Charts 107
Root-Cause Analysis 107
Conventions for Construction of Graphs 109
Sequential Frequency Chart Construction 110

SECTION 6 — Correlation and Regression

Correlation 117
Correlation Coefficient 117
Prediction 120
Regression 121
Standard Error of Estimate 124

SECTION 7 — Sampling and Estimation

Sampling Error 131
Sampling Distribution for Means 131
Sampling Distribution for Proportions 134

Confidence Limits . . . 135
Confidence Limits for Means of Large Samples . . . 136
Confidence Limits for Means of Small Samples . . . 136
Confidence Limits for Proportions . . . 139
Normal Distribution . . . 139
Binomial Distribution . . . 141
Poisson Distribution . . . 141

SECTION 8 — Significance Testing

Tests of Significance . . . 153
Directional and Non-directional Tests . . . 154
Tests of Significance for Means: t Distribution . . . 154
Independent Samples . . . 156
Correlated Samples . . . 158
Unequal Population Variance . . . 160
Tests of Significance for Proportions: Z Distribution . . . 162
Independent Samples . . . 162
Correlated Samples . . . 164
Tests of Significance for Frequencies: χ^2 Distribution . . . 166
Independent Samples . . . 167
Correlated Samples . . . 168
Same-Sample . . . 170
Test of Significance for Variance: F Distribution . . . 171
Independent Samples . . . 171
Correlated Samples . . . 173
Determining Sample Sizes . . . 175
Sample Size for Means . . . 175
Sample Size for Proportions . . . 176

SECTION 9 — Design of Experiments

Experimental Designs . . . 209
Independent and Dependent Variables . . . 209
Randomization . . . 210
Single Factor Experiments . . . 211
Analysis of Variance: One-Way Classification . . . 211
Degrees of Freedom . . . 212
Variance Estimates . . . 212
Factorial Experiments . . . 216
Interaction . . . 218
Analysis of Variance: Two-Way Classification . . . 222
Interaction . . . 231
Nonparametric Designs . . . 232

Analysis of Variance: Nonparametric 238
Replicated Runs 241

Bibliography 265

Index 267

TABLES

2.1 Poisson Distribution Curves 18
3.1 Table of z Values 39
4.1 Multiplication Factors for X and R Control Limits 51
7.1 z Values for Confidence Limits 135
7.2 t Values for Confidence Limits 138
7.3 Binomial Curves for Confidence Limits 142
7.4 Poisson Curves for Confidence Limits 143
8.1 z Values for Significance Testing 164
8.2 Critical Values of t 202
8.3 Critical Values of Chi Square (χ^2) 203
8.4 Critical Values of F: 5% level 204
8.5 Critical Values of F: 10% level 205

PREFACE

To succeed in a world economy, we need to design, develop, and produce superior products in minimum time periods with maximum returns on investment. Surprisingly, few engineering and manufacturing professionals are aware of the statistical tools that are available to help them meet these challenges.

Why are they unaware of these tools? One reason is that engineering schools don't always teach statistical methods and when they do, courses may be theoretical rather than practical. Textbooks often present formulas instead of explanations. Their authors may fail to convey information in terms the reader can readily grasp.

Similarly, working professionals who instruct others may compound the misconception that statistics is an unusually difficult subject. And, an instructor who hasn't mastered advanced statistical techniques may only be capable of teaching rudimentary methods.

The primary goal of this book is to relate statistics to real-world engineering and manufacturing problems, and to do so in a manner that anyone with some recollection of high school algebra can understand. I have tried to avoid abstract mathematics, to clarify the meanings of necessary symbols and terms, and to reconstruct formulas starting with their root forms. This book will move you forward, step-by-step, from basic concepts to practical applications of sophisticated and valuable statistical techniques.

A further goal of this book is to establish an understanding of statistical concepts that will enhance the way we manage ourselves and others; that we will be more likely to approach problems in a systematic manner and to base decisions on the analysis of data rather than on opinion or emotion.

J. E. Brown

ABSTRACT

This book is designed to provide engineering and manufacturing practitioners with a working understanding of the various statistical techniques and tools applicable in designing and producing reliable products.

Section 1
Frequency Distributions and Probability Curves

Acquaints the student with the basic structure of frequency distribution and associated theoretical probability curves. It establishes the foundation necessary in understanding applied statistical methods and is prerequisite to the remaining sections.

Section 2
Attribute Sampling

Develops an understanding of the risks associated with acceptance sampling and provides students with the knowledge and skills necessary to compute sampling plans which minimize those risks.

Section 3
Normal Distributions

Covers measures of central tendency, variance, and how the normal probability curve can be used to determine process capability. This section also forms the basis for understanding statistical control charts.

Section 4
Statistical Control Charts

Discusses the theory of all normally used control charts, the machine/process actions necessary before control charts are used, and probabilities associated with significant patterns or runs.

Section 5
Data Gathering, Presentation, and Basic Problem Solving

Details data gathering requirements in manufacturing, how to compile and display data for management reporting, and the quantitative approach to basic problem solving.

Section 6
Correlation and Regression

Covers the fundamental concepts of correlation studies, the pitfalls of commonly performed regression analysis, and techniques for establishing supplier/equipment guard-bands.

Section 7
Sampling and Estimation

Provides a more in-depth analysis of probability and sampling distributions. It covers the quantification of error associated with product/process predictions and the methods for establishing confidence limits.

Section 8
Significance Testing

Provides an in-depth study of methods for determining the degree of difference between sample parameters. It is application-oriented by covering the methods for determining if new processes, products, test procedures, etc., are significantly different from the ones they're replacing, or if environmental conditions, burn-in procedures, etc., have a significant effect on product performance.

Section 9
Design of Experiments

Acquaints the reader with the techniques and procedures for performing efficient and economical product performance and yield improvement experiments by analyzing the effects and interactions of multiple manipulated product/process variables.

SECTION 1 

FREQUENCY DISTRIBUTIONS AND PROBABILITY CURVES

DESCRIPTIVE AND INFERENTIAL STATISTICS

Descriptive Statistics are those which summarize or describe the key characteristics of known data, i.e., the salary of 500 employees. Since a list of 500 numbers is difficult to interpret, an average and range value, or some other similar statistic usually is computed (computational method). Another approach is to represent the data in the form of a graph.

For example, using the computational method (the average and range) we would say that the average annual salary for the 500 employees is $32,000 with a range of $15,000 to $60,000.

Using the graphical approach, we would construct a frequency distribution, or some form thereof, and visually interpret the data (Figure 1.1).

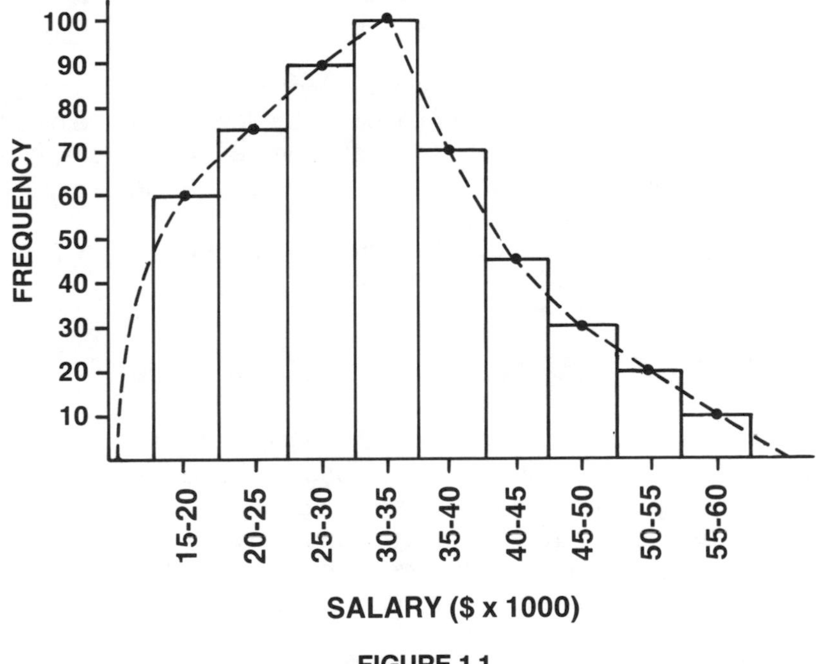

FIGURE 1.1

This gives a little more information of how the salaries are distributed, but what if these 500 data points merely represented a random sample drawn from a population of 10,000 employees, or from the entire American labor force? What could we say about the average salary for the entire population, or what percentage is above or below a certain salary point?

Inferential Statistics go beyond mere description. They provide a means of making inferences about populations from sample data. Before accurate inferences can be made, however, we must have some understanding of the shape of the distribution from which the sample was drawn.

DEFINING FREQUENCY DISTRIBUTIONS

Since frequency distributions differ with different sets of data, it is important to define the characteristics which describe how they differ from one another.

Central Tendency: Refers to the measured value near the center of the distribution — the middle point (Figure 1.2).

Variance: Refers to the extent which values cluster around the central value. If all values are close to the central value, the variance is less than if they tend to depart more markedly from the central value (Figure 1.3).

Skewness: Refers to the symmetry or asymmetry of the frequency distribution. If the majority of values tend to concentrate toward the low end of the distribution, the distribution is positively skewed. If they concentrate toward the high end, the distribution is negatively skewed (Figure 1.4).

Kurtosis: Refers to the flatness or peakedness of a distribution in relation to the "normal" distribution (Figure 1.5).

PROBABILITY CURVES

In statistical work, the most important characteristic of a frequency distribution is its shape, or profile. However, since most distributions of actual observed data tend to be somewhat irregular, precise inference about populations is often impossible. For this reason, based on the assumption that the curves would be uniform if the entire population was accounted for, several mathematically based theoretical distributions are used in making population inferences.

By assuming that the population is uniform, the frequency distribution also becomes a probability distribution. For example, if a given proportion of a frequency distribution contains 1/6 of the total area under the curve, the probability of drawing a sample from that area is 1/6 or 0.16. Accordingly, the probability of not drawing a sample from that area is 5/6 or 0.83.

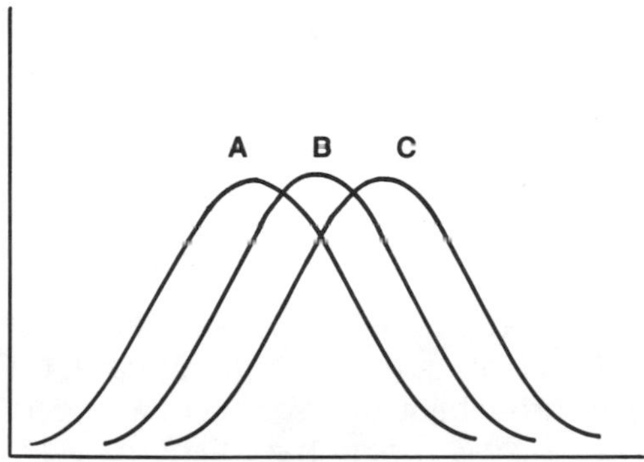

Three frequency distributions with the same variance but with different central points.

FIGURE 1.2

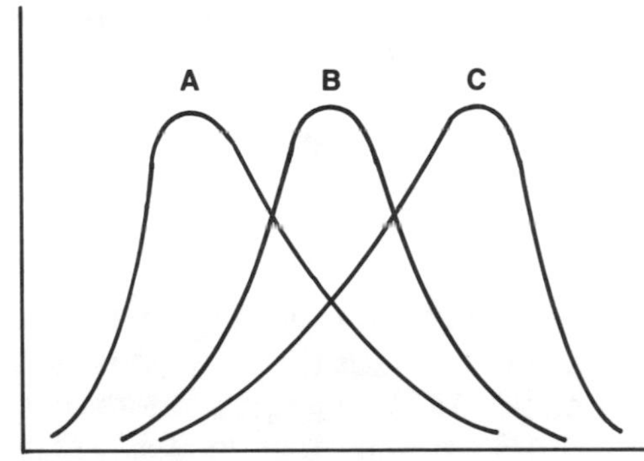

Three frequency distributions differing in skewness.

FIGURE 1.4

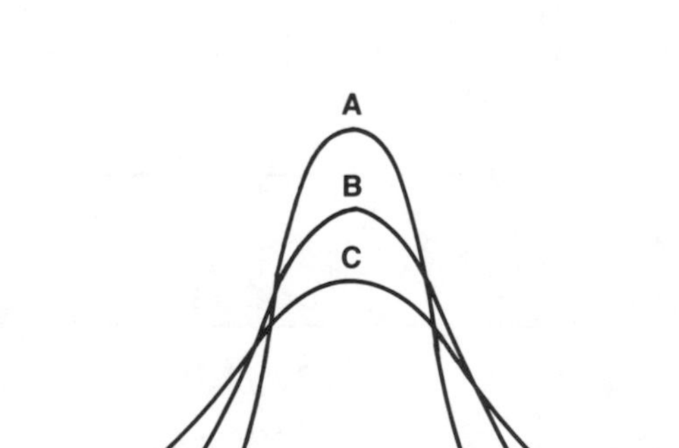

Three frequency distributions with the same central point but with different variance.

FIGURE 1.3

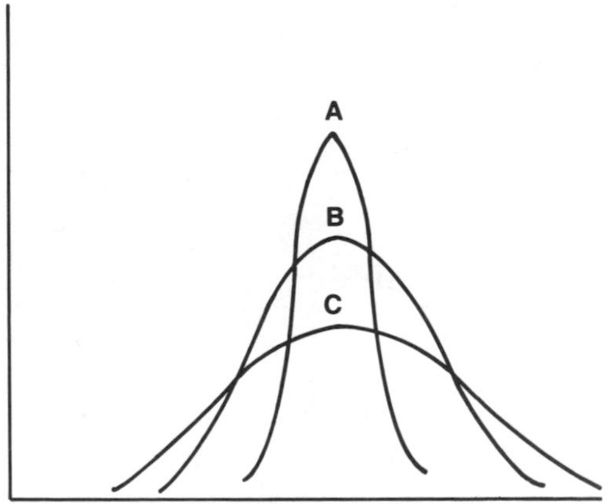

Three frequency distributions differing in kurtosis.

FIGURE 1.5

In applied statistics, the most commonly used theoretical distributions are the *binomial, poisson, F, t, chi square* (χ^2), and, of particular importance, the *normal* distribution (which will be discussed later).

BINOMIAL DISTRIBUTION

The binomial distribution is a theoretical probability distribution where a different curve exists for different combinations of acceptable events, unacceptable events, and sample sizes. For example, the shape of the distribution is determined by the equation $(p + q)^n$ where p is the probability that an event will occur, q is the probability that it will not occur, and n is the number of observations (i.e., the sample size).

The binomial expansion $(p + q)^n$ is readily illustrated in solving the probabilities of obtaining five, four, three, two, one, and zero 6's in rolling five dice. In this case, the probability of obtaining a 6 on a roll of a die, the probability that the event will occur (p), is 1/6. Likewise, the probability of not obtaining a 6, the probability of the event not occurring (q), is 5/6.

The probabilities associated with all five dice are then:

$$(p + q)^n \text{ or } (1/6 + 5/6)^5$$

which is:

$$(1/6)^5 + 5(1/6)^4(5/6) + 10(1/6)^3(5/6)^2 + 10\,(1/6)^2(5/6)^3 + 5(1/6)(5/6)^4 + (5/6)^5$$

Number of 6's	Binomial Term	Probability
5	$(1/6)^5$	.000128
4	$5(1/6)^4(5/6)$	.003215
3	$10(1/6)^3(5/6)^2$	.032150
2	$10(1/6)^2(5/6)^3$	.160750
1	$5(1/6)(5/6)^4$	.401877
0	$(5/6)^5$	.401877

As seen, this distribution is asymmetrical.

If the value of p and q are 1/2, or .5, as in tossing five coins, the probability distribution is symmetrical (i.e., .031 for five heads, .156 for four heads, .312 for three heads, .312 for two heads, .156 for one head, and .031 for zero heads).

In this case, when the distribution is symmetrical, the mean will be equal to n/2, the variance will be equal to n/4, skewness will be equal to 0, and kurtosis will be equal to − 1/n.

When the values of p and q depart from .5, the distribution will depart from symmetry, becoming more and more skewed as p or q becomes smaller and smaller (Figure 1.6). For this reason, since a different curve exists for differing values of p and q, the mean, variance, skewness, and kurtosis of any binomial distribution is defined as:

$$\text{Mean} = np \text{ (where n is the sample size)}$$

$$\text{Variance} = npq$$

$$\text{Skewness} = \frac{p\text{-}q}{\sqrt{npq}}$$

$$\text{Kurtosis} = \frac{1 - 6pq}{npq}$$

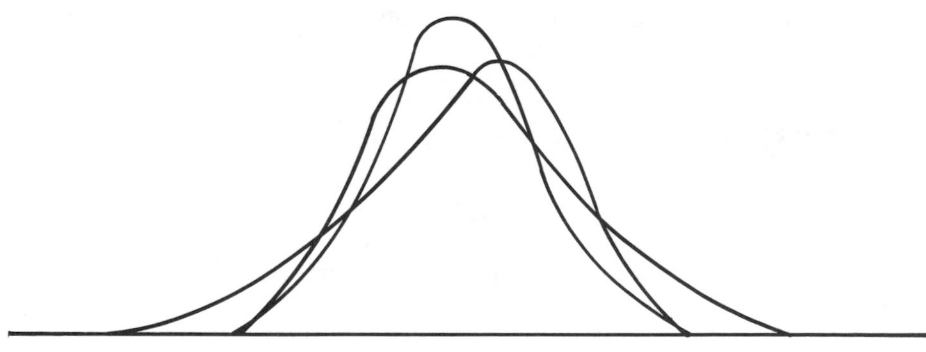

FIGURE 1.6

POISSON DISTRIBUTION

The poisson distribution is similar to the binomial in that it consists of a number of terms which, respectively, give the probability of 0, 1, 2, 3, or more occurrences for a given population average and sample size.

While the shape of the binomial and poisson distributions are essentially identical for any given set of conditions, the poisson distribution is more manageable for large sample quantities. The reason being that the binomial distribution is defined if n and p are known, while the poisson is defined if the product of n and p are known. In practical terms, two parameters are required for the binomial while one is required for the poisson (Figure 1.7).

FIGURE 1.7

Since the poisson distribution(s) is defined by the product of np, the mean and standard deviation for any particular curve is defined by the equations: Mean = np and standard deviation = $\sqrt{np}$.

t DISTRIBUTION

The t distribution, unlike the binomial and poisson distributions, is always symmetrical about the mean. The exact shape of the distribution, however, is dependent on the associated degrees of freedom (df), which in most instances is equal to the sample size minus 1 ($n - 1$). When the sample size is small, the extremities are thicker and more spread than for larger sample sizes. For each curve, therefore, the associated probability values are different (Figure 1.8).

For degrees of freedom exceeding 30 ($n \geq 31$), the t distribution closely approximates the normal distribution and, at that point, the probability values remain fairly constant.

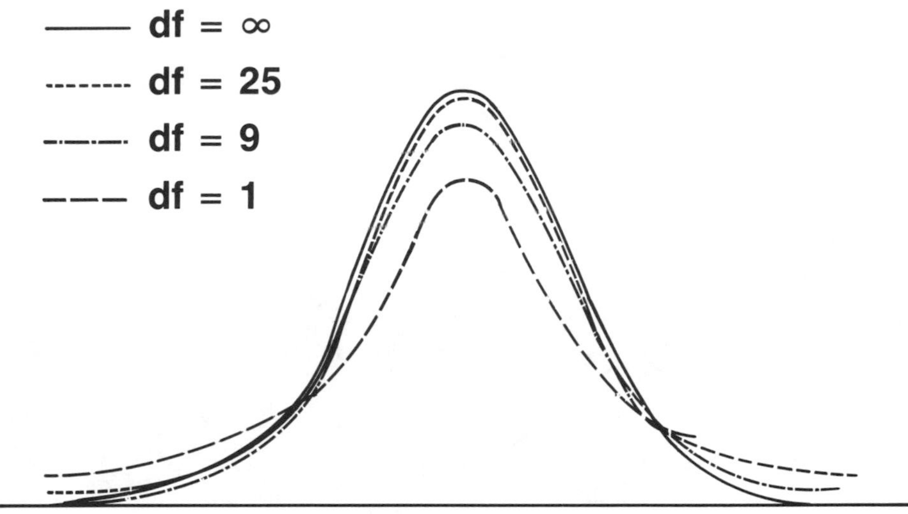

FIGURE 1.8

CHI SQUARE (χ^2) DISTRIBUTION

The chi square (χ^2) distribution is of considerable importance because it provides a method of comparing the frequency of observed data to expected theoretical frequencies. In this manner, it is possible to test the "goodness of fit" of observed data to the various theoretical distributions and, based on the results, select the appropriate distribution for data analysis and prediction.

In addition to this useful application, the χ^2 distribution provides a ready means for testing significance of difference between both proportions and sample variances.

Like the t and F distributions, while not always symmetrical, the shape of the χ^2 distribution is determined by the degrees of freedom (Figure 1.9).

F DISTRIBUTION

The F distribution is defined as a probability distribution of sample variance ratios. It provides a means for comparing the relationship of two independent sample variances, the ratio between the two, to determine if the samples are significantly different (i.e., if the two samples represent the same or different populations).

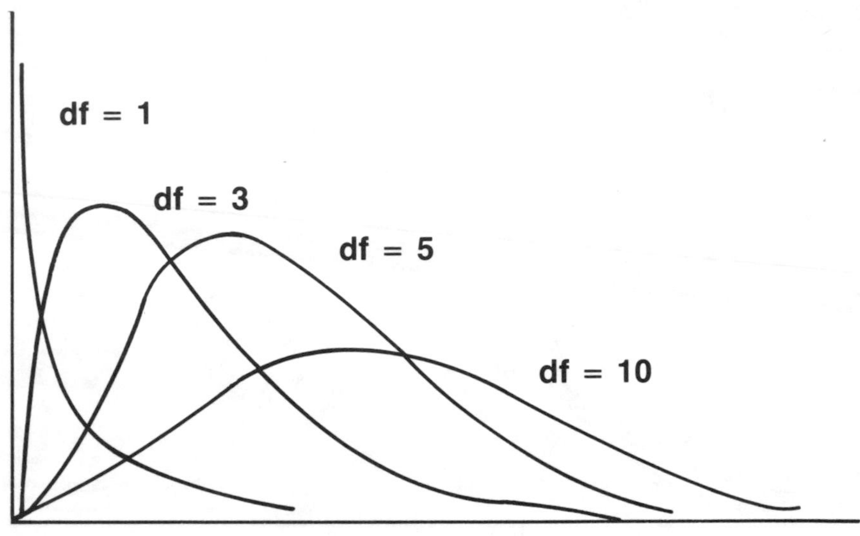

FIGURE 1.9

Three important characteristics of the F distribution are that all values comprising the distribution are positive (equal to or greater than 0), that the distribution is always skewed to the right, and that a different distribution exists for every combination of degrees of freedom (Figure 1.10).

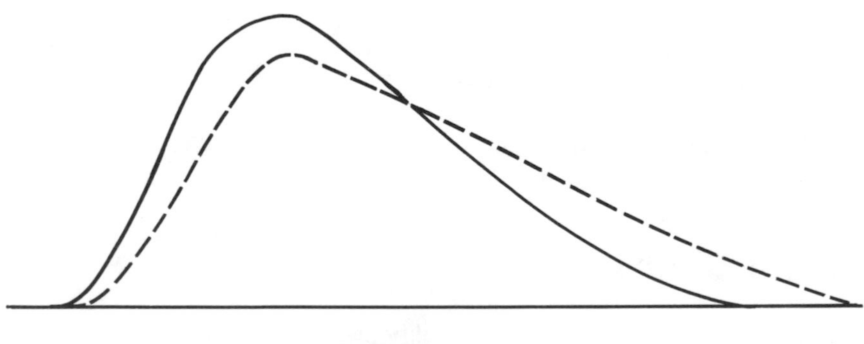

FIGURE 1.10

SECTION 2

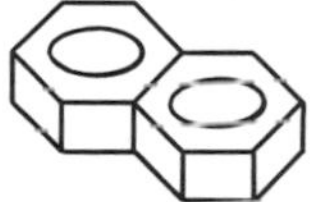

ATTRIBUTE SAMPLING

ATTRIBUTE SAMPLING DEFINITION

Attribute: A characteristic or property which is appraised in terms of whether it does or does not meet specified requirements.

Sample: A product or quantity of products chosen at random to represent all products in a batch or lot.

Attribute Sampling: Measurement of quality by the method of attributes consists of verifying the conformance or nonconformance of each sample in a sample group and, based on the number of nonconforming samples, determining whether the lot should be accepted or rejected.

What else is attribute sampling? A game of chance.

Attribute sampling is, for a given sample size and lot percent defective, the probability of not detecting a defective.

BINOMIAL SAMPLING PROBABILITIES

The binomial distribution (binomial expansion) generally is used to calculate sampling probabilities for small sample quantities.

The equation for the binomial expansion is:

$$(q + p)^n$$

Where (q) is the fractional proportion of acceptable items in a batch or lot, where (p) is the fractional proportion of defective items, and (n) is the sample size.

For a sample size of 2 from a lot that is (q) percent good and (p) percent bad, we have:

Possible Combinations	Way of Occurring	Probability of Each Way	Probability of Type of Arrangement
2 Good	**Good, Good**	qq	q^2
1 Good	**Good, Bad**	qp	$2qp$
1 Bad	**Bad, Good**	pq	
2 Bad	**Bad, Bad**	pp	p^2

By adding all the entries in the column labeled "probability of type of arrangement" we have:

$$q^2 + 2qp + p^2$$

where q^2 is the probability of no defectives (good, good), $2qp$ is the probability of one defective (good, bad or bad, good) and p^2 is the probability of two defectives (bad, bad).

Thus, for a sample of 2, the binomial expansion is:

$$(q + p)^2 = q^2 + 2qp + p^2$$

where $q + p$ for any value of n is always equal to 1:

$$(q + p)^n = 1.0$$

BINOMIAL SAMPLING PLANS

Three values are needed to calculate the binomial sampling probabilities: the fractional proportion of acceptable items in a lot or batch (q), the fractional proportion of defective items (p), and the sample size (n).

Example 2.1
For a sample size of 4 drawn from a lot that is 10% defective, what is the probability of observing 0, 1, 2, 3, or 4 defectives?

DEFECTIVES: [0] [1] [2] [3] [4]

$$(q + p)^4 = [(q)^4] + [4(q)^3(p)] + [6(q)^2(p)^2] + [4(q)(p)^3] + [(p)^4]$$

$$= [(.9)^4] + [4(.9)^3(.1)] + [6(.9)^2(.1)^2] + [4(.9)(.1)^3] + [(.1)^4]$$

$$= [.6561] + [.2916] + [.0486] + [.0036] + [.0001]$$

$$= 1.000$$

For lot acceptance sampling, since we're not interested in all possible probabilities, only the first few terms are normally used (i.e., the probability of 0 defectives, the probability of 1 defective, or the probability of 2 defectives).

Thus, in Example 2.1, for a sample size of 4 with an acceptance number of 0, there is a 0.6561 probability (65.61% chance) of accepting a lot that is 10% defective. Likewise, there is a 94.77% chance of accepting the lot if the acceptance number is 1 (i.e., the probability of 0 defectives [65.61%] plus the probability of 1 defective [29.16%]).

OPERATING CHARACTERISTIC CURVES

Four values are required to compute an operating characteristic curve. The sample size (n), the acceptance number (c), the fraction of acceptable items (q), and the fraction defective (p).

Sample Size

Example 2.2
For sample sizes of 10, 20, and 50, with an acceptance number of 0, calculate the probability of acceptance (p_a) associated with the lot quality levels listed.

Note: Since the acceptance number is 0 (the probability of observing 0 or less defectives), only the first term of the binomial and the lot fraction acceptable is needed. For ease in plotting the operating characteristic (OC) curves (Figure 2.1), however, the fraction defective is also listed.

Acceptance Number

Example 2.3
For a sample size of 50 with acceptance numbers of 1 and 2, calculate the probability of acceptance (Figure 2.2) associated with the lot quality levels listed.

Note: For an acceptance number of 1, the probability of acceptance is equal to the probability of 0 plus the probability of 1 (the sum of the first and second terms):

$$[(q)^{50} + 50(q)^{49}(p)]$$

For an acceptance number of 2, the probability of acceptance is equal to the sum of the first three terms:

$$[(q)^{50} + 50(q)^{49}(p) + 1225(q)^{48}(p)^2]$$

Lot Fraction Acc. (q)	def (p)	Sample :10 Acc. No: 0 $(q)^n$	p_a	Sample :20 Acc. No: 0 $(q)^n$	p_a	Sample :50 Acc. No: 0 $(q)^n$	p_a
.999	.001	$(.999)^{10}$	.99	$(.999)^{20}$	.98	$(.999)^{50}$	.95
.995	.005	$(.995)^{10}$	.95	$(.995)^{20}$	.90	$(.995)^{50}$	.77
.99	.01	$(.99)^{10}$	.90	$(.99)^{20}$	.81	$(.99)^{50}$	.60
.95	.05	$(.95)^{10}$	.59	$(.95)^{20}$	.35	$(.95)^{50}$	.07
.90	.10	$(.90)^{10}$	.34	$(.90)^{20}$	.12	$(.90)^{50}$	.005

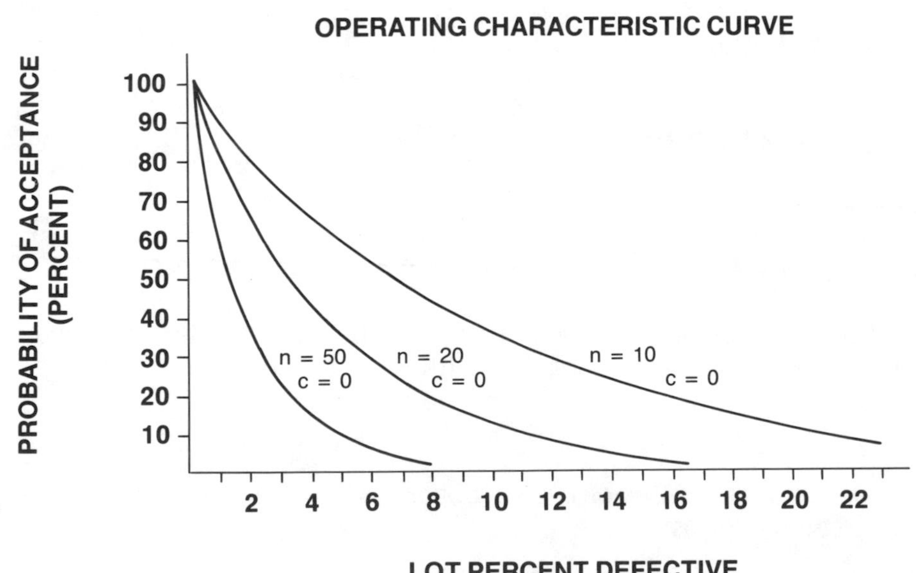

FIGURE 2.1

Lot Fraction acc.(q)	Lot Fraction def.(p)	Binomial Expansion For One Defective $(q)^{50} + 50(q)^{49}(p)$	p_a	Binomial Expansion For Two Defectives $(q)^{50} + 50(q)^{49}(p) + 1225(q)^{48}(p)^2$	p_a
.999	.001	$(.999)^{50} + 50(.999)^{49}(.001)$	.998	$(.999)^{50} + 50(.999)^{49}(.001) + 1225(.999)^{48}(.001)^2$	.999
.995	.005	$(.995)^{50} + 50(.995)^{49}(.005)$	.97	$(.995)^{50} + 50(.995)^{49}(.005) + 1225(.995)^{48}(.005)^2$	.997
.99	.01	$(.99)^{50} + 50(.99)^{49}(.01)$	.91	$(.99)^{50} + 50(.99)^{49}(.01) + 1225(.99)^{48}(.01)^2$	.98
.95	.05	$(.95)^{50} + 50(.95)^{49}(.05)$	.27	$(.95)^{50} + 50(.95)^{49}(.05) + 1225(.95)^{48}(.05)^2$	.54
.90	.10	$(.90)^{50} + 50(.90)^{49}(.10)$	.03	$(.90)^{50} + 50(.90)^{49}(.10) + 1225(.90)^{48}(.10)^2$	.11

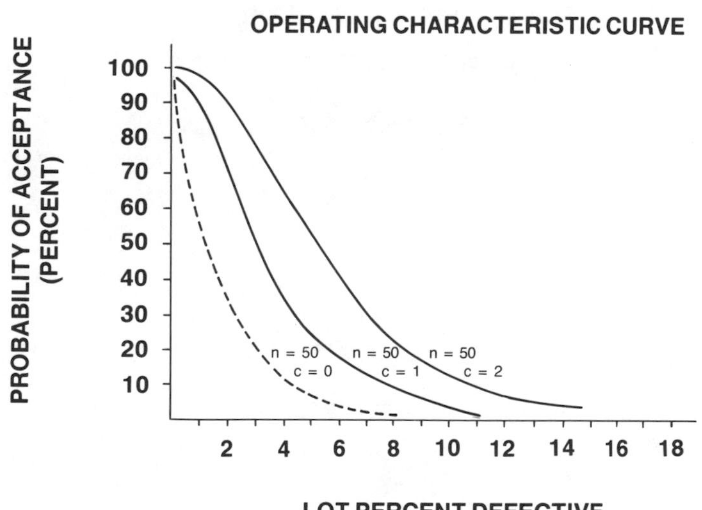

FIGURE 2.2

POISSON SAMPLING PLANS

For large sample sizes, considerable numerical computation is involved in using the binomial expansion. If the population fraction defective (p) is less than 0.1 however, regardless of the value of np (the sample size times the fraction defective), the poisson distribution provides a good approximation to the binomial distribution.

Example 2.4
For a sample size of 50 with acceptance numbers of 0 and 1, determine the probability of acceptance for a lot that is 10% defective.

$$n = 50, \quad p = 0.1, \quad np = 5$$

Read along the bottom of the poisson distribution curve (Table 2.1) to np of 5, then up to the curve marked 0. Then go to the left and read that the probability of 0 or less is approximately .0065. Thus, for an acceptance number of 0, the probability of acceptance is .0065 or 0.65%.

For an acceptance number of 1, read across to np of 5, then up to the curve marked 1. Then go to the left and read that the probability of acceptance is approximately .04 or 4%.

TABLE 2.1 Poisson Distribution Curves

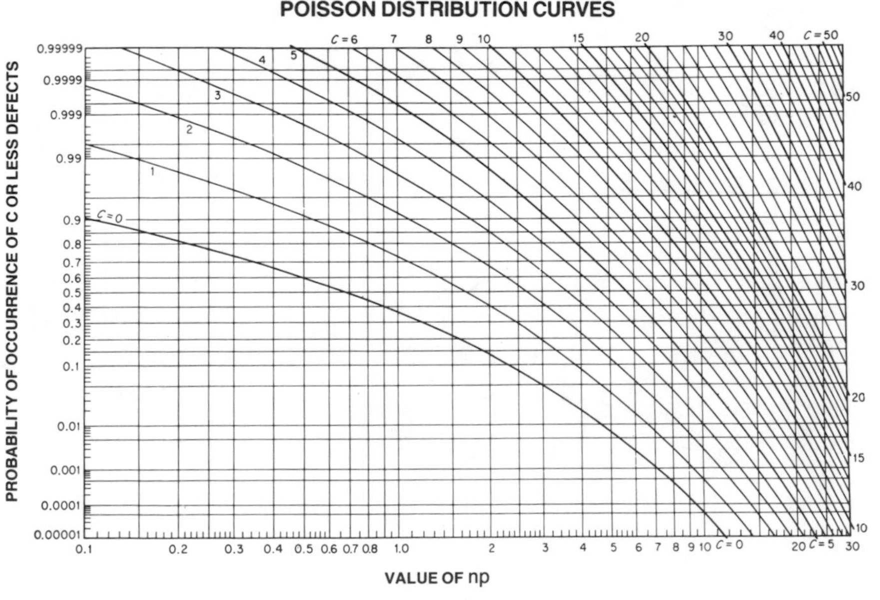

Example 2.5
For a sample size of 50 with acceptance numbers of 0, 1, and 2, determine the probability of acceptance associated with the following fraction defective values:

.005, .01, .05, .10, .15

For a fraction defective value of .005, read along the bottom of the poisson distribution curve (Table 2.1) to np of .25. Then up to the curve marked 0. Then go to the left and read that the probability of 0 or less is approximately .78.

Return to np of .25 and go up to the curve marked 1. Go to the left and read that the probability of 1 or less is approximately .975.

Return to np of .25 and go up to the curve marked 2. Go to the left and read that the probability of 2 or less is approximately .998.

Perform the same actions for the remaining np values. Then, using the resulting probabilities of acceptance for the listed lot fraction defective values (converted to lot percent defective), plot the operating characteristic curves (Figure 2.3).

SAMPLING RISKS

All sampling plans involve risk. There is the risk of rejecting "good" lots, as well as the risk of accepting "bad" lots.

Acceptable Quality Level (AQL): The percentage of defective items that is considered acceptable.

Reject Quality Level (RQL): The percentage of defective items that is considered unacceptable.

Producers Risk: The probability of rejecting lots that are equal to or better than the specified acceptable quality level. This is commonly referred to as the alpha (α), or type 1, risk.

Consumers Risk: The probability of accepting lots that are equal to or worse than the specified reject quality level. This is commonly referred to as the beta (β), or type 2, risk.

Fraction Defective (p)	Sample Size (n)	(np)	Probability of Acceptance (c = 0)	(c = 1)	(c = 2)
.005	50	.25	.78	.975	.998
.01	50	.5	.61	.910	.986
.05	50	2.5	.08	.285	.545
.10	50	5.0	.0065	.040	.16
.15	50	7.5	.0005	.005	.02

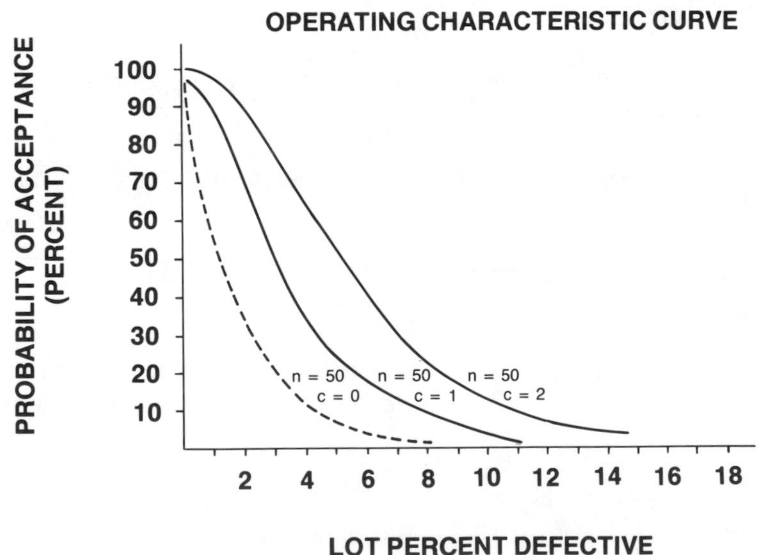

FIGURE 2.3

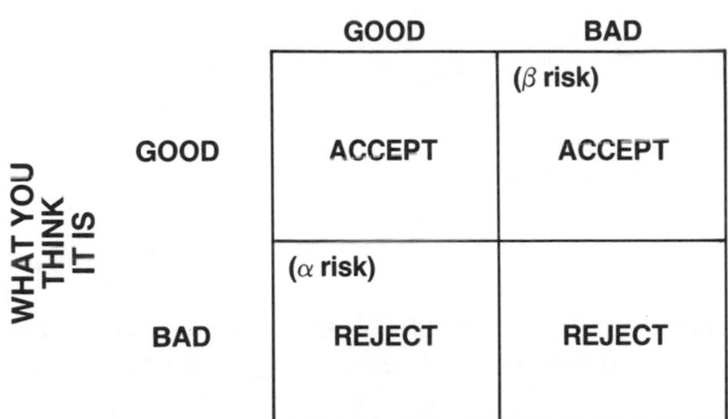

WHAT IT REALLY IS
GOOD
BAD
WHAT YOU THINK IT IS
GOOD
BAD
(β risk)
ACCEPT
ACCEPT
(α risk)
REJECT
REJECT

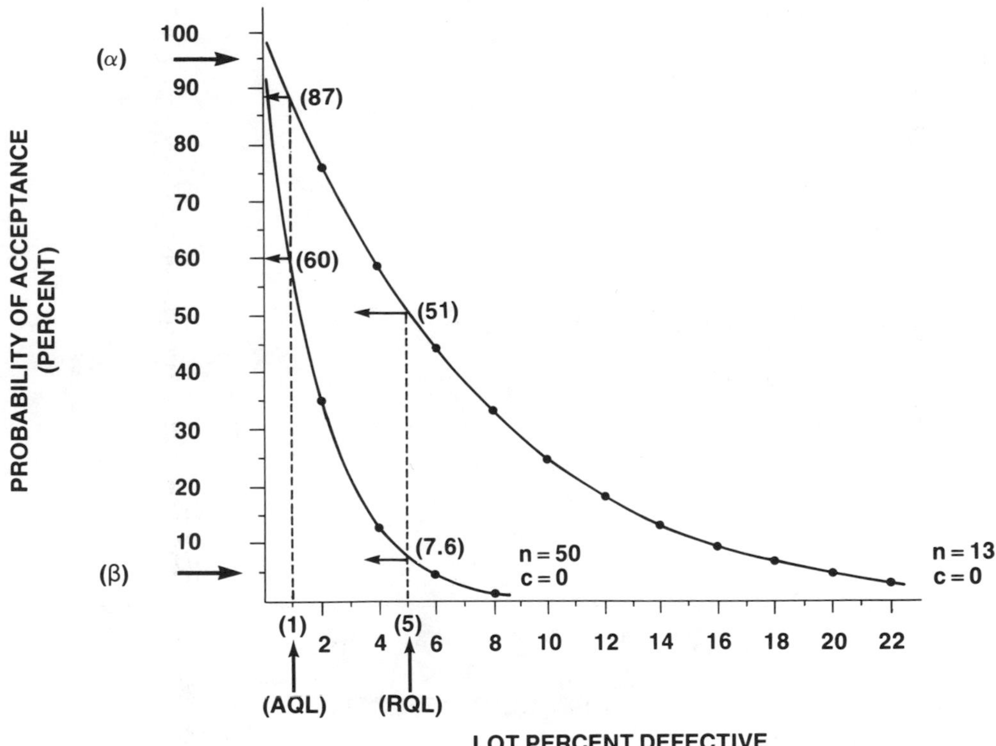

PROBABILITY OF ACCEPTANCE (PERCENT)
100
90
80
70
60
50
40
30
20
10
(α)
(β)
(87)
(60)
(51)
(7.6)
n = 50
c = 0
n = 13
c = 0
(1)
2
4
(5)
6
8
10
12
14
16
18
20
22
(AQL)
(RQL)
LOT PERCENT DEFECTIVE

FIGURE 2.4

Exercise Worksheets

Exercise 2.1

OPERATING CHARACTERISTIC CURVES
(BINOMIAL DISTRIBUTION)

For the lot percent defective values listed on the binomial worksheet, calculate the probability of acceptance (Pa) for a sample size of 10 and 20 with acceptance numbers of 0. When completed, plot the resulting probabilities on the graph below.

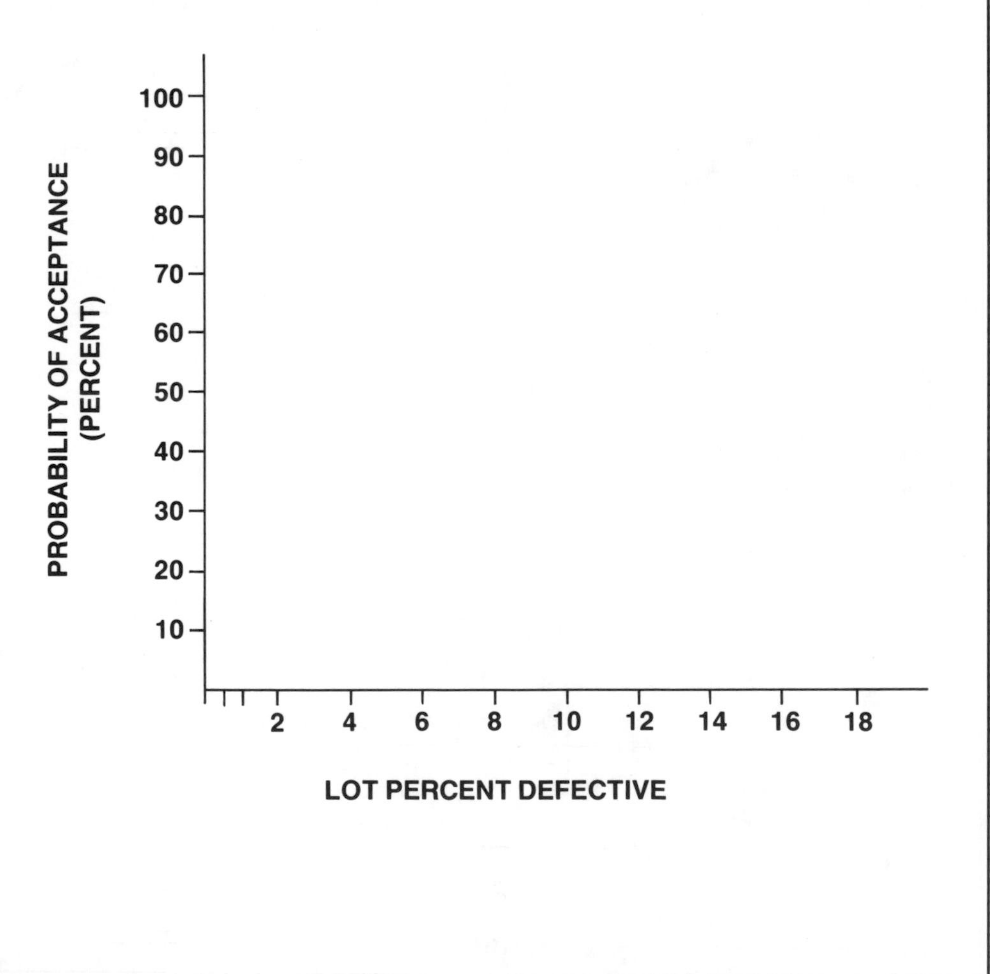

Exercise 2.1 (continued)

BINOMIAL WORKSHEET

Lot % Def.	Lot Fraction Acc. (q)	Lot Fraction Def. (p)	Sample Size : 10 Acceptance No. : 0 Term: $(q)^n$	Pa	Sample size : 20 Acceptance No. : 0 Term: $(q)^n$	Pa
0.5%	[ ]	[ ]	________	= [ ]	________	= [ ]
1.0%	[ ]	[ ]	________	= [ ]	________	= [ ]
2.0%	[ ]	[ ]	________	= [ ]	________	= [ ]
4.0%	[ ]	[ ]	________	= [ ]	________	= [ ]
6.0%	[ ]	[ ]	________	= [ ]	________	= [ ]
8.0%	[ ]	[ ]	________	= [ ]	________	= [ ]
12.0%	[ ]	[ ]	________	= [ ]	________	= [ ]
16.0%	[ ]	[ ]	________	= [ ]	________	= [ ]

Exercise 2.2

OPERATING CHARACTERISTIC CURVES
(POISSON DISTRIBUTION)

For the lot percent defective values listed on the poisson worksheet, determine the probability of acceptance (Pa) for a sample size of 100 with acceptance numbers of 0 and 1. When completed, plot the resulting probabilities on the graph below.

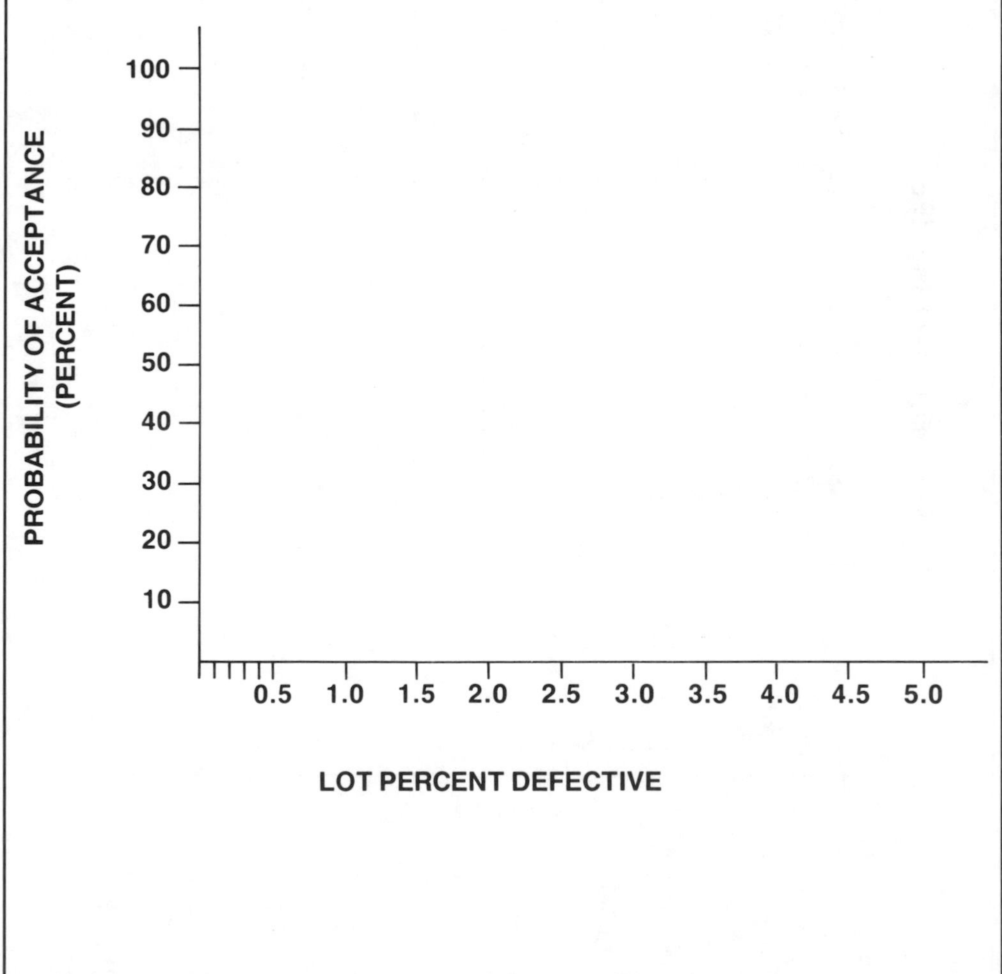

Exercise 2.2 (continued)

POISSON WORKSHEET

Lot Percent Defective	Fraction Defective (p)	Sample Size (n)	(np)	Probability of Acceptance	
				Acc. No. : 0	Acc. No. : 1
0.1%	.001	100	[ ]	[ ]	[ ]
0.3%	.003	100	[ ]	[ ]	[ ]
0.5%	.005	100	[ ]	[ ]	[ ]
1.0%	.010	100	[ ]	[ ]	[ ]
2.0%	.020	100	[ ]	[ ]	[ ]
3.0%	.030	100	[ ]	[ ]	[ ]
4.0%	.040	100	[ ]	[ ]	[ ]
5.0%	.050	100	[ ]	[ ]	[ ]

SECTION 3

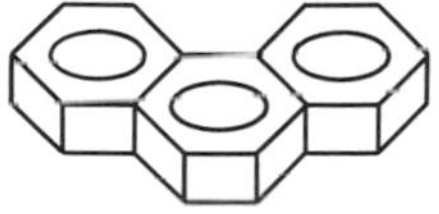

NORMAL DISTRIBUTIONS

THE NORMAL CURVE

The most obvious characteristic of the normal distribution is its bell-shaped curve. This bell shape results from the manner in which values randomly disperse themselves around the mean due to chance alone. Most values occur close to the mean with fewer and fewer values occurring as they move toward the tails of the distribution (Figure 3.1).

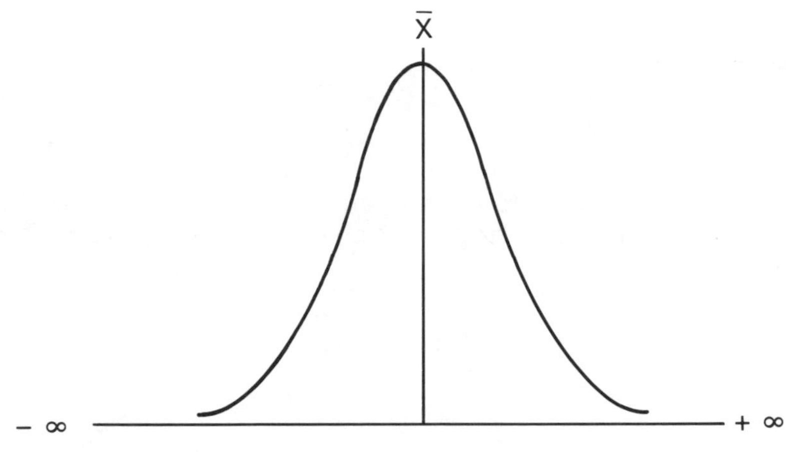

FIGURE 3.1

MEASURES OF CENTRAL TENDENCY

Mean: The average value of all measurements in a set of data.

$$\text{Mean } (\bar{x}) = \Sigma\, x/n$$
$$\bar{x} = 8 + 6 + 10 + 5 + 11 = 40/5 = 8$$

Median: The value, in order of magnitude, in which half the values are above and half are below.

Values = 2, 7, 16, 19, 20, 25, 27
Median = 19

Mode: The most frequently occurring value.

Values = 11, 11, 12, 12, 12, 13, 13, 13, 13, 13, 14, 14, 14, 15, 15, 15, 16, 16, 17, 17, 18

Mode = 13

MEASURES OF DISPERSION

Range: The difference between the lowest and highest values in a set of data.

$$\text{Range (R)} = x_H - x_L$$
Values 8, 6, 10, 5, 11
$$R = 11 - 5 = 6$$

Variance: The average of the summed squared values in which each value varies from the average of all values.

$$s^2 = \frac{\Sigma (x - \bar{x})^2}{n - 1}$$

Standard Deviation: The square root of the sample variance (i.e., the square root of the squared and then averaged value by which each data point in the distribution varies from the mean).

$$s = \sqrt{\frac{\Sigma (x - \bar{x})^2}{n - 1}}$$

MOMENTS ABOUT THE MEAN

$$m_1 = \frac{\Sigma (x - \bar{x})}{n}$$

$$m_2 = \frac{\Sigma (x - \bar{x})^2}{n}$$

$$m_3 = \frac{\Sigma (x - \bar{x})^3}{n}$$

$$m_4 = \frac{\Sigma (x - \bar{x})^4}{n}$$

Skewness: The sum of cubes of deviation about the mean.

$$\text{Skewness } (g_1) = \frac{m_3}{m_2 \sqrt{m_2}}$$

When the distribution is symmetrical, the sum of the cubes of deviations above the mean will balance the sum of cubes below the mean. Thus, when $m_3 = 0$, $g_1 = 0$. With a tail to the right, g_1 is positive, with a tail to the left, g_1 is negative.

Kurtosis: The measure of flatness and/or peakedness in relation to the normal distribution.

$$\text{Kurtosis } (g_2) = \frac{m_4}{(m_2)^2} - 3$$

When g_2 is 0 the distribution is normal, flatter than the normal when g_2 is negative, and more peaked than normal when g_2 is positive.

For central tendency of a distribution, the mean is usually the preferred measure.

For dispersion, the standard deviation is the most commonly applied statistical measure.

CALCULATING STANDARD DEVIATION

Six Steps:

1) Calculate the average ($\bar{x}$) for the data set.
2) Subtract the average from each individual value of (x). Note: Some results will be positive and some negative.
3) Square each result.
4) Add the squared results together.
5) Divide that total by the total number of values minus 1 (n − 1).
6) Find the square root of that value.

x value	$\bar{x}$	$x - \bar{x}$	$(x - \bar{x})^2$
$x_1 = 5$	6	−1	1
$x_2 = 7$	6	1	1
$x_3 = 6$	6	0	0
$x_4 = 8$	6	2	4
$x_5 = 4$	6	−2	4
$x_6 = 8$	6	2	4
$x_7 = 7$	6	1	1
$x_8 = 4$	6	−2	4
$x_9 = 5$	6	−1	1
$x_{10} = 6$	6	0	0
$\Sigma x = 60$	$\frac{\Sigma x}{n} = 6$		$\Sigma(x - \bar{x})^2 = 20$

$$s = \sqrt{\frac{\Sigma\,(x - \bar{x})^2}{n - 1}} = \sqrt{\frac{20}{9}} = 1.49$$

AREA UNDER THE NORMAL CURVE

The inherent shape of the normal distribution dictates that 99.73% of all values, or area under the curve, occur within ±3 standard deviations from the mean. About 95% (95.45%) occur within ±2 standard deviations, and 68% (68.26%) occur within ±1 standard deviation (Figure 3.2).

In this manner, the area under the curve can be equally divided into 6 standard deviation units: 3 to the right or "plus" side of the mean, and 3 units to the left or "minus" side of the mean.

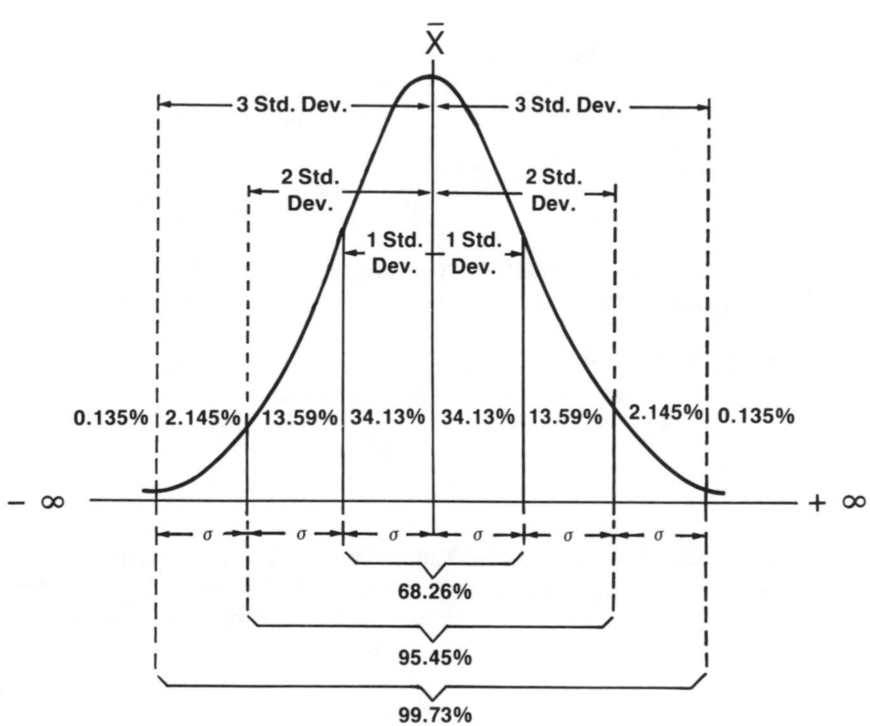
X̄
3 Std. Dev.
3 Std. Dev.
2 Std. Dev.
2 Std. Dev.
1 Std. Dev.
1 Std. Dev.
0.135%
2.145%
13.59%
34.13%
34.13%
13.59%
2.145%
0.135%
− ∞
+ ∞
σ
σ
σ
σ
σ
σ
68.26%
95.45%
99.73%

FIGURE 3.2

Sample A		Sample B	
$x_1 = 18$	$x_6 = 20$	$x_1 = 8$	$x_6 = 9$
$x_2 = 20$	$x_7 = 22$	$x_2 = 9$	$x_7 = 5$
$x_3 = 21$	$x_8 = 21$	$x_3 = 9$	$x_8 = 7$
$x_4 = 19$	$x_9 = 18$	$x_4 = 7$	$x_9 = 10$
$x_5 = 22$	$x_{10} = 19$	$x_5 = 8$	$x_{10} = 8$
MEAN = 20.0		**MEAN = 8.0**	
STD. DEV. = 1.49		**STD. DEV. = 1.41**	

For Sample A: 99.73% of all values occur between 15.5 and 24.4, 95.45% occur between 17.0 and 22.9, and 68.26% occur between 18.5 and 21.4.

For Sample B: 99.73% of all values occur between 3.7 and 12.2, 95.45% occur between 5.1 and 10.8, and 68.26% occur between 6.5 and 9.4.

THE z SCORE

The z score provides a means for determining various areas under the curve that are not exactly 1, 2, or 3 standard deviations away from the mean. With the aid of a z table (at the end of this section), it is possible to determine any area proportion between $\pm$ 3.59 standard deviations from the mean.

The z score formula converts the mean of the distribution from its actual value to 0. Distance from the mean is then measured, not in terms of actual values, but in terms of the number of standard deviations away from the mean.

$$z = \frac{x - \bar{x}}{s}$$

Example 3.1
For a distribution of values with a mean of 30 and a standard deviation of 5, what percentage of values fall below 22?

$$z = \frac{x - \bar{x}}{s} = \frac{22 - 30}{5} = -1.6$$

Turning to the z table for a value of − 1.6, we find that 0.0548 or 5.48% of the area under the curve lies between minus infinity and − 1.6 standard deviations from the mean. Thus, 5.48% of the distribution values fall below 22.

Example 3.2

For a distribution of values with a mean of 34 and a standard deviation of 3, what percentage of values fall outside the range of 30 to 40?

Below 30 $$z = \frac{x - \bar{x}}{s} = \frac{30 - 34}{3} = -1.33$$

$$z \text{ of } -1.33 = 0.0918 = 9.18\%$$

Above 40 $$z = \frac{x - \bar{x}}{s} = \frac{40 - 34}{3} = +2.0$$

$$z \text{ of } +2.0 = 0.9773 = 97.73\%$$

$$100\% - 97.73\% = 2.27\%$$

Below 30 and Above 40

$$9.18\% + 2.27\% = 11.45\%$$

Exercise Worksheets

Exercise 3.1

STANDARD DEVIATION

WORKSHEET

X VALUE	$\bar{X}$	$X-\bar{X}$	$(X-\bar{X})^2$
$X_1 = 33$	______	______	______
$X_2 = 31$	______	______	______
$X_3 = 29$	______	______	______
$X_4 = 31$	______	______	______
$X_5 = 33$	______	______	______
$X_6 = 36$	______	______	______
$X_7 = 34$	______	______	______
$X_8 = 30$	______	______	______
$X_9 = 28$	______	______	______
$X_{10} = 33$	______	______	______

$\Sigma X = [\quad]$ $\quad \frac{\Sigma X}{n} = [\quad]$ $\quad \Sigma(X-\bar{X})^2 = [\quad]$

$$s = \sqrt{\frac{\Sigma(X-\bar{X})^2}{n-1}} = \sqrt{\frac{[\quad]}{[\quad]}} = [\quad]$$

Exercise 3.2

AREA UNDER THE NORMAL CURVE

(WORKSHEET 1)

X_1 = 41.0	X_{11} = 38.6	X_{21} = 39.3
X_2 = 39.9	X_{12} = 40.1	X_{22} = 39.8
X_3 = 40.8	X_{13} = 39.1	X_{23} = 40.0
X_4 = 39.0	X_{14} = 39.6	X_{24} = 38.5
X_5 = 40.3	X_{15} = 40.6	X_{25} = 38.3
X_6 = 40.2	X_{16} = 39.6	X_{26} = 40.5
X_7 = 40.3	X_{17} = 39.2	X_{27} = 40.4
X_8 = 39.1	X_{18} = 40.4	X_{28} = 39.2
X_9 = 39.2	X_{19} = 39.0	X_{29} = 39.0
X_{10} = 39.0	X_{20} = 40.8	X_{30} = 38.9

Mean = [ ] **Std. Dev.** = [ ]

99.73% will fall between [ ] **and** [ ]

95.45% will fall between [ ] **and** [ ]

68.26% will fall between [ ] **and** [ ]

Exercise 3.3

AREA UNDER THE NORMAL CURVE

(WORKSHEET 2)

For a distribution of parts with a mean of 4 inches and a standard deviation of 0.5 inches, what percentage of the parts will fall: (a) below 3.2 inches, (b) above 4.7 inches, and (c) between 3.2 and 4.7 inches

(a) below: [ ]

(b) above: [ ]

(c) between: [ ]

TABLE 3.1 Table of z Values

Area Under the Normal Distribution Curve Between ± 3.59 Standard Deviations from the Mean.

Z VALUE	0.09	0.08	0.07	0.06	0.05	0.04	0.03	0.02	0.01	0.00
−3.5	0.00017	0.00017	0.00018	0.00019	0.00019	0.00020	0.00021	0.00022	0.00022	0.00023
−3.4	0.00024	0.00025	0.00026	0.00027	0.00028	0.00029	0.00030	0.00031	0.00033	0.00034
−3.3	0.00035	0.00036	0.00038	0.00039	0.00040	0.00042	0.00043	0.00045	0.00047	0.00048
−3.2	0.00050	0.00052	0.00054	0.00056	0.00058	0.00060	0.00062	0.00064	0.00066	0.00069
−3.1	0.00071	0.00074	0.00076	0.00079	0.00082	0.00085	0.00087	0.00090	0.00094	0.00097
−3.0	0.00100	0.00104	0.00107	0.00111	0.00114	0.00118	0.00122	0.00126	0.00131	0.00135
−2.9	0.0014	0.0014	0.0015	0.0015	0.0016	0.0016	0.0017	0.0017	0.0018	0.0019
−2.8	0.0019	0.0020	0.0021	0.0021	0.0022	0.0023	0.0023	0.0024	0.0025	0.0026
−2.7	0.0026	0.0027	0.0028	0.0029	0.0030	0.0031	0.0032	0.0033	0.0034	0.0035
−2.6	0.0036	0.0037	0.0038	0.0039	0.0040	0.0041	0.0043	0.0044	0.0045	0.0047
−2.5	0.0048	0.0049	0.0051	0.0052	0.0054	0.0055	0.0057	0.0059	0.0060	0.0062
−2.4	0.0064	0.0066	0.0068	0.0069	0.0071	0.0073	0.0075	0.0078	0.0080	0.0082
−2.3	0.0084	0.0087	0.0089	0.0091	0.0094	0.0096	0.0099	0.0102	0.0104	0.0107
−2.2	0.0110	0.0113	0.0116	0.0119	0.0122	0.0125	0.0129	0.0132	0.0136	0.0139
−2.1	0.0143	0.0146	0.0150	0.0154	0.0158	0.0162	0.0166	0.0170	0.0174	0.0179
−2.0	0.0183	0.0188	0.0192	0.0197	0.0202	0.0207	0.0212	0.0217	0.0222	0.0228
−1.9	0.0233	0.0239	0.0244	0.0250	0.0256	0.0262	0.0268	0.0274	0.0281	0.0287
−1.8	0.0294	0.0301	0.0307	0.0314	0.0322	0.0329	0.0336	0.0344	0.0351	0.0359
−1.7	0.0367	0.0375	0.0384	0.0392	0.0401	0.0409	0.0418	0.0427	0.0436	0.0446
−1.6	0.0455	0.0465	0.0475	0.0485	0.0495	0.0505	0.0516	0.0526	0.0537	0.0548
−1.5	0.0559	0.0571	0.0582	0.0594	0.0606	0.0618	0.0630	0.0643	0.0655	0.0668
−1.4	0.0681	0.0694	0.0708	0.0721	0.0735	0.0749	0.0764	0.0778	0.0793	0.0808
−1.3	0.0823	0.0838	0.0853	0.0869	0.0885	0.0901	0.0918	0.0934	0.0951	0.0968
−1.2	0.0985	0.1003	0.1020	0.1038	0.1057	0.1075	0.1093	0.1112	0.1131	0.1151
−1.1	0.1170	0.1190	0.1210	0.1230	0.1251	0.1271	0.1292	0.1314	0.1335	0.1357
−1.0	0.1379	0.1401	0.1423	0.1446	0.1469	0.1492	0.1515	0.1539	0.1562	0.1587
−0.9	0.1611	0.1635	0.1660	0.1685	0.1711	0.1736	0.1762	0.1788	0.1814	0.1841
−0.8	0.1867	0.1894	0.1922	0.1949	0.1977	0.2005	0.2033	0.2061	0.2090	0.2119
−0.7	0.2148	0.2177	0.2207	0.2236	0.2266	0.2297	0.2327	0.2358	0.2389	0.2420
−0.6	0.2451	0.2483	0.2514	0.2546	0.2578	0.2611	0.2643	0.2676	0.2709	0.2743
−0.5	0.2776	0.2810	0.2843	0.2877	0.2912	0.2946	0.2981	0.3015	0.3050	0.3085
−0.4	0.3121	0.3156	0.3192	0.3228	0.3264	0.3300	0.3336	0.3372	0.3409	0.3446
−0.3	0.3483	0.3520	0.3557	0.3594	0.3632	0.3669	0.3707	0.3745	0.3783	0.3821
−0.2	0.3859	0.3897	0.3936	0.3974	0.4013	0.4052	0.4090	0.4129	0.4168	0.4207
−0.1	0.4247	0.4286	0.4325	0.4364	0.4404	0.4443	0.4483	0.4522	0.4562	0.4602
−0.0	0.4641	0.4681	0.4721	0.4761	0.4801	0.4840	0.4880	0.4920	0.4960	0.5000

TABLE 3.1 (continued)

Z VALUE	0.00	0.01	0.02	0.03	0.04	0.05	0.06	0.07	0.08	0.09
+0.0	0.5000	0.5040	0.5080	0.5120	0.5160	0.5199	0.5239	0.5279	0.5319	0.5359
+0.1	0.5398	0.5438	0.5478	0.5517	0.5557	0.5596	0.5636	0.5675	0.5714	0.5753
+0.2	0.5793	0.5832	0.5871	0.5910	0.5948	0.5987	0.6026	0.6064	0.6103	0.6141
+0.3	0.6179	0.6217	0.6255	0.6293	0.6331	0.6368	0.6406	0.6443	0.6480	0.6517
+0.4	0.6554	0.6591	0.6628	0.6664	0.6700	0.6736	0.6772	0.6808	0.6844	0.6879
+0.5	0.6915	0.6950	0.6985	0.7019	0.7054	0.7088	0.7123	0.7157	0.7190	0.7224
+0.6	0.7257	0.7291	0.7324	0.7357	0.7389	0.7422	0.7454	0.7486	0.7517	0.7549
+0.7	0.7580	0.7611	0.7642	0.7673	0.7704	0.7734	0.7764	0.7794	0.7823	0.7852
+0.8	0.7881	0.7910	0.7939	0.7967	0.7995	0.8023	0.8051	0.8079	0.8106	0.8133
+0.9	0.8159	0.8186	0.8212	0.8238	0.8264	0.8289	0.8315	0.8340	0.8365	0.8389
+1.0	0.8413	0.8438	0.8461	0.8485	0.8508	0.8531	0.8554	0.8577	0.8599	0.8621
+1.1	0.8643	0.8665	0.8686	0.8708	0.8729	0.8749	0.8770	0.8790	0.8810	0.8830
+1.2	0.8849	0.8869	0.8888	0.8907	0.8925	0.8944	0.8962	0.8980	0.8997	0.9015
+1.3	0.9032	0.9049	0.9066	0.9082	0.9099	0.9115	0.9131	0.9147	0.9162	0.9177
+1.4	0.9192	0.9207	0.9222	0.9236	0.9251	0.9265	0.9279	0.9292	0.9306	0.9319
+1.5	0.9332	0.9345	0.9357	0.9370	0.9382	0.9394	0.9406	0.9418	0.9429	0.9441
+1.6	0.9452	0.9463	0.9474	0.9484	0.9495	0.9505	0.9515	0.9525	0.9535	0.9545
+1.7	0.9554	0.9564	0.9573	0.9582	0.9591	0.9599	0.9608	0.9616	0.9625	0.9633
+1.8	0.9641	0.9649	0.9656	0.9664	0.9671	0.9678	0.9686	0.9693	0.9699	0.9706
+1.9	0.9713	0.9719	0.9726	0.9732	0.9738	0.9744	0.9750	0.9756	0.9761	0.9767
+2.0	0.9773	0.9778	0.9783	0.9788	0.9793	0.9798	0.9803	0.9808	0.9812	0.9817
+2.1	0.9821	0.9826	0.9830	0.9834	0.9838	0.9842	0.9846	0.9850	0.9854	0.9857
+2.2	0.9861	0.9864	0.9868	0.9871	0.9875	0.9878	0.9881	0.9884	0.9887	0.9800
+2.3	0.9893	0.9896	0.9898	0.9901	0.9904	0.9906	0.9909	0.9911	0.9913	0.9916
+2.4	0.9918	0.9920	0.9922	0.9925	0.9927	0.9929	0.9931	0.9932	0.9934	0.9936
+2.5	0.9938	0.9940	0.9941	0.9943	0.9945	0.9946	0.9948	0.9949	0.9951	0.9952
+2.6	0.9953	0.9955	0.9956	0.9957	0.9959	0.9960	0.9961	0.9962	0.9963	0.9964
+2.7	0.9965	0.9966	0.9967	0.9968	0.9969	0.9970	0.9971	0.9972	0.9973	0.9974
+2.8	0.9974	0.9975	0.9976	0.9977	0.9977	0.9978	0.9979	0.9979	0.9980	0.9981
+2.9	0.9981	0.9982	0.9983	0.9983	0.9984	0.9984	0.9985	0.9985	0.9986	0.9986
+3.0	0.99865	0.99869	0.99874	0.99878	0.99882	0.99886	0.99889	0.99893	0.99896	0.99900
+3.1	0.99903	0.99906	0.99910	0.99913	0.99915	0.99918	0.99921	0.99924	0.99926	0.99929
+3.2	0.99931	0.99934	0.99936	0.99938	0.99940	0.99942	0.99944	0.99946	0.99948	0.99950
+3.3	0.99952	0.99953	0.99955	0.99957	0.99958	0.99960	0.99961	0.99962	0.99964	0.99965
+3.4	0.99966	0.99967	0.99969	0.99970	0.99971	0.99972	0.99973	0.99974	0.99975	0.99976
+3.5	0.99977	0.99978	0.99978	0.99979	0.99980	0.99981	0.99981	0.99982	0.99983	0.99983

SECTION 4

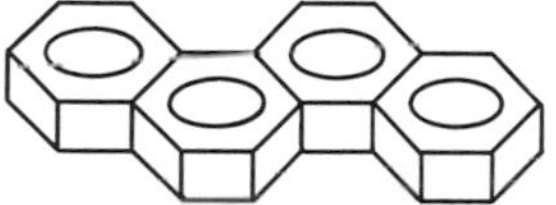

STATISTICAL CONTROL CHARTS

DETECTION VERSUS PREVENTION

Traditionally, the control of quality involved the classification of products as either acceptable or not acceptable after they were produced. In contrast, control charts are used to judge the quality of products as they are being produced. This is accomplished by periodically drawing random samples from the process and determining, based on one or more sample statistics, if the process is producing products within acceptable statistical limits.

This is the fundamental difference between inspection sampling and statistical control charts. Inspection sampling tells us something about the quality of completed products. Statistical control charts tell us something about the variability of products as they're being produced.

PROCESS CAPABILITY

An effective procedure for assuring continuous production of quality products is to perform a time-sequenced analysis of sample parts as they're actually being produced. In this manner, it is possible to not only verify that the process is capable of producing parts within specification limits, but ascertain whether it can do so on a continuous basis (i.e., maintain proper adjustment over time).

Following this procedure, the first step is to adjust the process to produce parts at the specification midpoint. This means adjusting the process so that the mean of the distribution of parts will occur midway between the upper and lower specification limits. Then, under actual operating conditions, a consecutive sample of at least 50 parts are drawn from the process, the applicable product characteristics measured, and the measurements recorded in chronological order.

The next step is to calculate the sample standard deviation and z distance between the specification midpoint and the upper and lower specification limits.

The specification midpoint value, instead of the sample mean, is used to eliminate the possibility of the capability study being influenced by improper process adjustments. For example, to assure that the process is consistently capable of producing acceptable parts, there should be at least 4 standard deviation units between the specification midpoint and the upper and lower specification limits. This allows the mean of the process to shift plus and minus 1 standard deviation around the specification midpoint without moving any closer than 3 standard deviations from the specification limits. This assures that 99.73% of the parts will always be within specification limits (Figure 4.1).

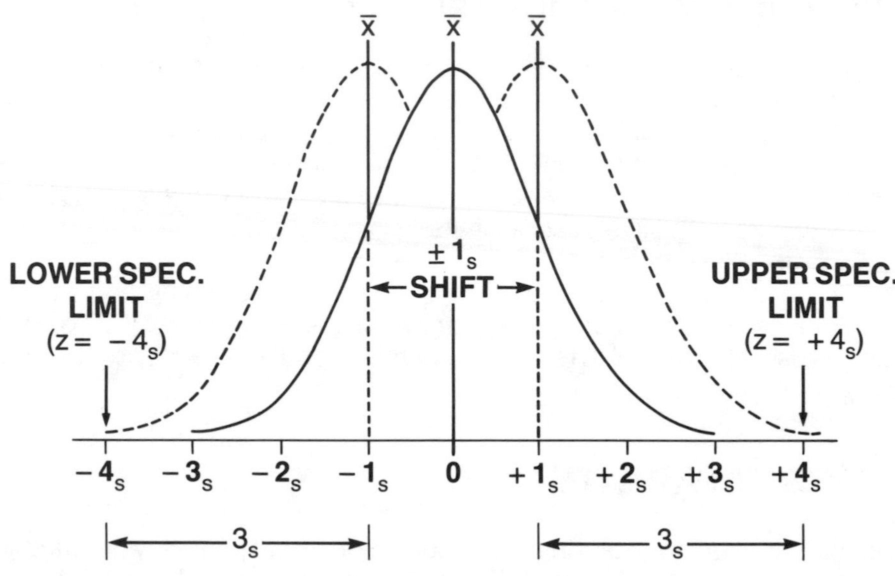

FIGURE 4.1

Example 4.1

Given the following 50 sample measurements, determine if the process is capable of producing products to a specification requirement of 40 $\pm$ 3 unit values.

CHRONOLOGICAL SAMPLE MEASUREMENTS				
39.2	38.6	40.4	39.6	40.0
39.5	40.5	39.8	39.2	39.9
39.6	39.1	40.6	40.4	40.8
40.5	40.5	39.0	39.0	38.5
40.8	40.8	39.3	40.8	40.3
40.0	38.8	39.1	40.0	40.8
39.5	40.7	40.4	40.4	40.3
40.3	39.3	39.7	40.8	39.6
39.6	39.7	40.9	38.7	39.4
40.6	41.1	39.5	39.2	39.0
$\bar{x} = 39.8$	s = .71			

$$\text{Upper Limit: } z = \frac{x - smp}{s} = \frac{43 - 40}{.71} = +\ 4.22$$

$$\text{Lower Limit: } z = \frac{x - smp}{s} = \frac{37 - 40}{.71} = -\ 4.22$$

Since the upper and lower specification limits are greater than 4 standard deviations from thc specification midpoint, the process is considered capable (i.e., all products will be within specification limits if the process mean is maintained within $\pm$ 1 standard deviation).

PROCESS STABILITY

For a process to lend itself to the application of control charts, it must be capable of producing products well within specification limits, and do so over a reasonable period of time.

To determine how long the process can operate without readjustment, the 50-piece sample is broken down into subsets of 5 samples each, the average for each subset calculated and then plotted on a run chart. A run chart is simply a graphical display of the subset averages over time (Figure 4.2).

If the process is subject to downward drift, for example, a suitable readjustment cycle must be established to prevent the manufacture of defective products.

This is accomplished by viewing the relationship between the lower specification limit and the sample standard deviation (i.e., by calculating the distance between the lower specification limit and the lowest possible process mean that would still provide for a displacement of 3 standard deviations):

$$x = z \times s + L_{SL}$$

where x is the lowest possible mean value, $z = 3$, s is the sample standard deviation, and L_{SL} is the lower specification limit (Figure 4.3).

SUBSET SAMPLING

As will be discussed in detail in Section 7, small samples provide little accuracy in predicting the true mean of a population or process. If a series of small samples are drawn from a process however, and the average of the sample averages computed, this grand average ($\bar{\bar{x}}$) will give an accurate prediction of the process mean (Figure 4.4).

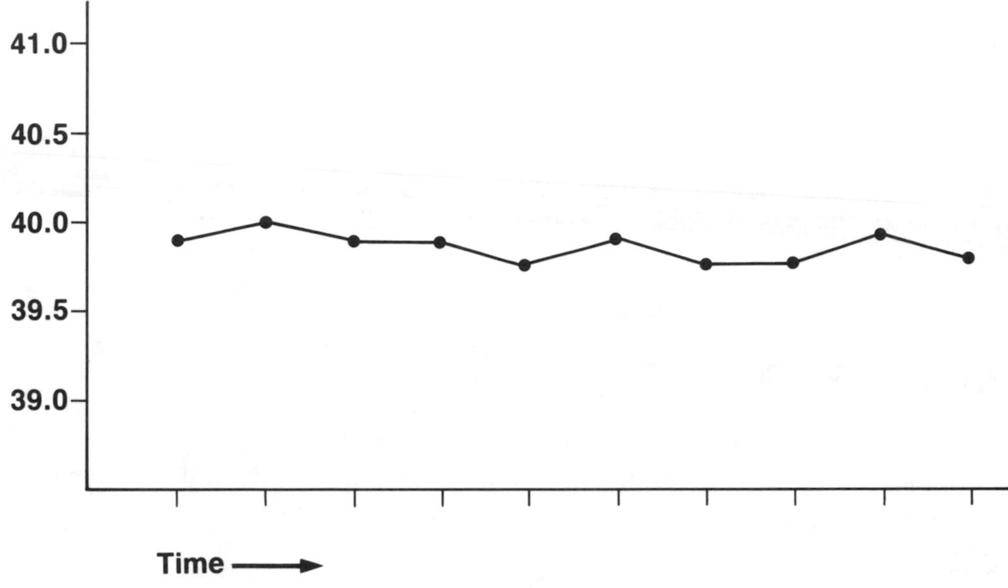

FIGURE 4.2

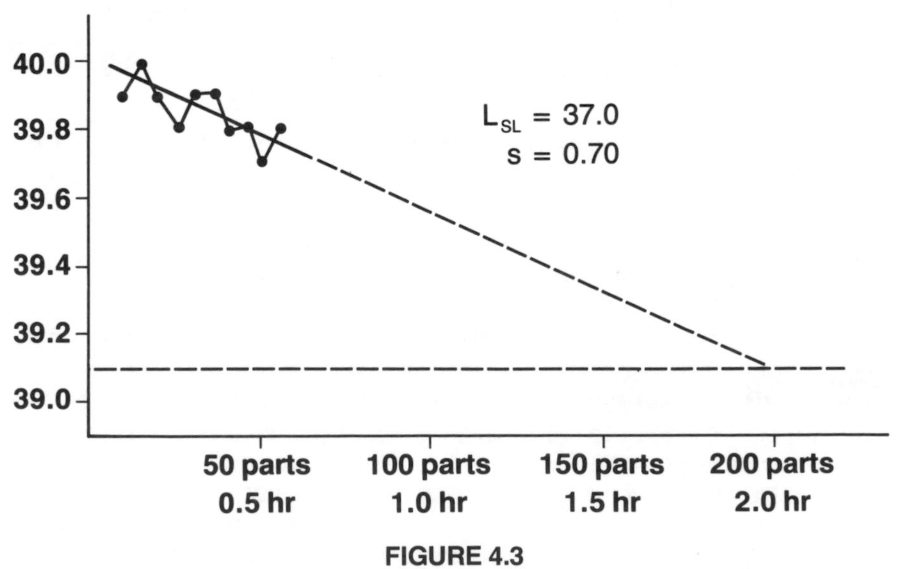

FIGURE 4.3

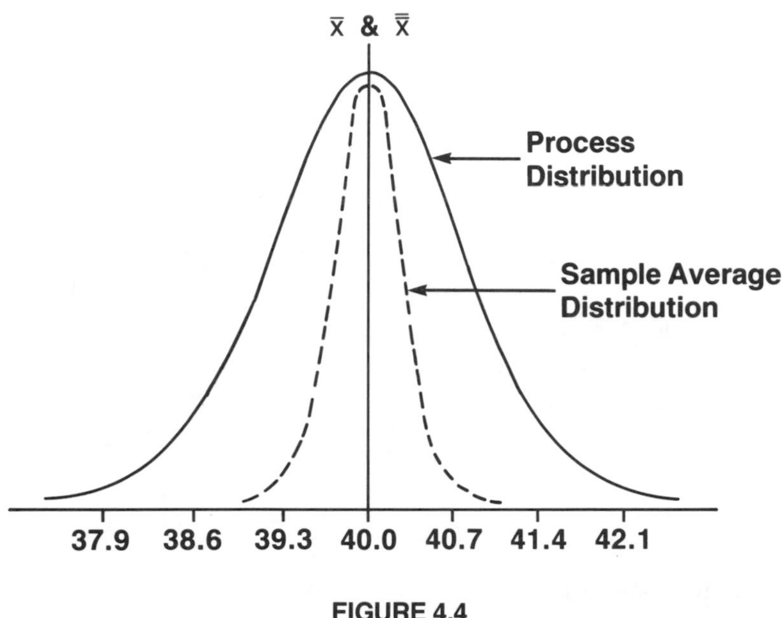

FIGURE 4.4

Since the samples are averaged, the standard deviation of the sample average distribution, or sampling distribution, is smaller than the process distribution. For a subset sample size of 5, the standard deviation of the sampling distribution is approximately .44 times the population, or process, distribution.

It is important to note that the sampling distribution, regardless of the shape of the population distribution, will always tend to be normal. For example, if the population distribution is square, there is an equal chance of drawing a single sample from anywhere within the distribution. Thus, when the samples are averaged, the average will always move toward the center.

The same is true for a triangular distribution. In this case, however, since there is a greater chance of drawing samples from the side of the distribution having the greater frequency of values, the mean of the sampling distribution will be closer to that side. But again, since the samples are averaged, the sampling distribution will tend to be normal (Figure 4.5).

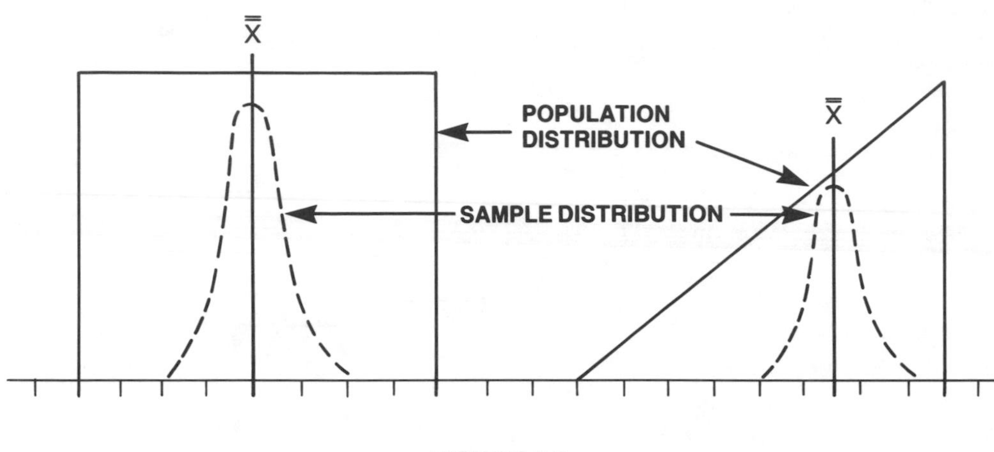

FIGURE 4.5

CONTROL CHART FOR AVERAGES

Since the sampling distribution follows the curve of a normal distribution, the same rules apply to this distribution as to any normal distribution. However, in this case, instead of working on the principle of 99.73% of all values occurring within $\pm$ 3 standard deviations from the population mean ($\bar{x}$), we now work on the principle that 99.73% of all sample averages will occur between $\pm$ 3 standard deviations of the sample average mean ($\bar{\bar{x}}$) (Figure 4.6).

Accordingly, since there is only a 0.27% chance of a sample average falling outside those limits due to chance alone, and if a sample average occurs in that area, then there is a 99.73% chance that the sample average mean has shifted. And if the sample average mean has shifted, so has the process mean.

This is the basic theory behind statistical control charts. A sample mean is viewed in relation to the probability of occurrence of a chance happening. If that probability is greater than would be expected due to chance alone, the process is deemed to be out-of-control.

The control chart monitors process shift. It does not assure 100% acceptable products unless the process itself is capable of producing products well within specification limits (Figure 4.7).

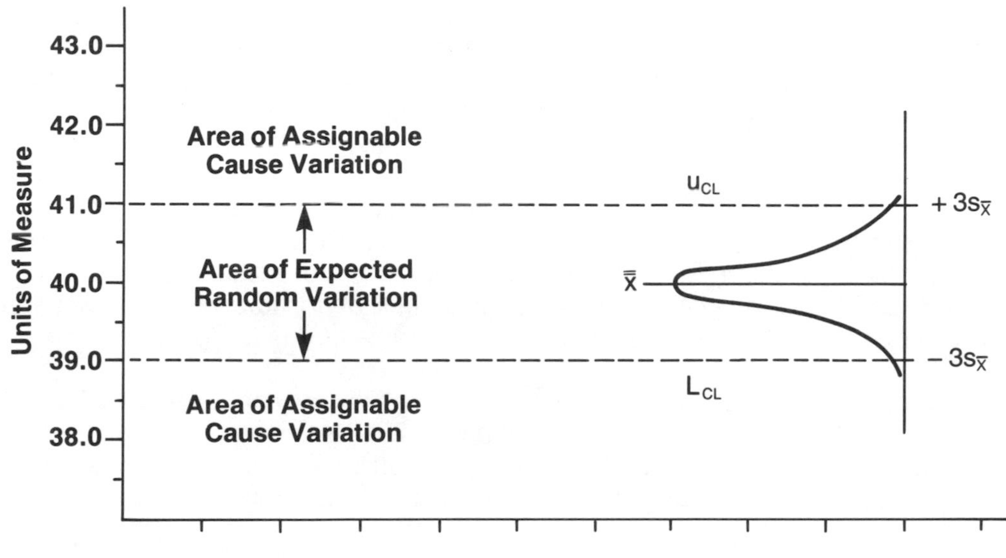

FIGURE 4.6

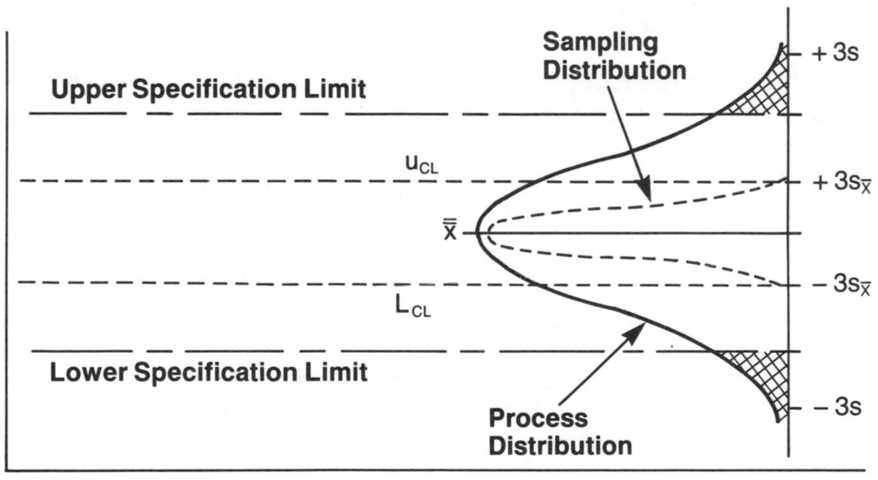

FIGURE 4.7

CONTROL CHART FOR RANGE

If the process being monitored exhibits relatively constant dispersion over time, then the range chart provides less useful information than the average chart. Where this is not the case, however, the range chart, in conjunction with the average chart, provides an effective means for monitoring both process variation and dispersion.

While the distribution of sample averages is normally distributed on both sides of the mean, the sample range distribution is skewed to the right, especially for small sample sizes. For this reason, the upper and lower range control limits are not at equal distances from the range mean ($\overline{R}$). In practical application, however, this makes little difference since the lower control limit for a sample of 5 is always 0.

For the upper range control limit, while the probability may not be exactly 99.87% at + 3 standard deviations, the general concepts are the same as the average chart. There is the area of normal random process dispersion (the area within the control limits), as well as the area of assignable cause dispersion (the area beyond the control limit).

CALCULATING CONTROL LIMITS

Since standard deviation calculations are tedious to perform, the following simplified methods are usually preferred:

Control Limits for Average: Control limits for the average chart are calculated using the formulas:

$$U_{CL} = SMP + A_2 \times \overline{R}$$

$$L_{CL} = SMP - A_2 \times \overline{R}$$

where SMP is the specification midpoint, A_2 is a constant multiplication factor associated with a given sample size, and $\overline{R}$ is the average of the range values (Table 4.1)

The control limit formula using SMP should be used when it is desirable to monitor and/or reflect process performance in relation to specification midpoint values. When this is not the case, or when this cannot be done, the $\overline{\overline{X}}$ value should be used.

TABLE 4.1 Multiplication Factors for X and R Control Limits

NUMBER OF SAMPLES IN SUBGROUP	FACTORS FOR X CHART	FACTORS FOR R CHART	
		Lower Control Limit	Upper Control Limit
	A_2	D_3	D_4
*1	2.66	—	—
2	1.88	0	3.27
3	1.02	0	2.57
4	0.73	0	2.28
5	0.58	0	2.11
6	0.48	0	2.00
7	0.42	0.08	1.92
8	0.37	0.14	1.86
9	0.34	0.18	1.82
10	0.31	0.22	1.78

*See page 58.

Control Limits for Range: The control limits for the range chart are computed using the formulas:

$$U_{CL} = D_4 \times \overline{R}$$
$$L_{CL} = D_3 \times \overline{R}$$

where D_3 and D_4 are constant multiplication factors associated with a given sample size.

Example 4.2

Given the following measurement information for 7 samples of 5 each, calculate the upper and lower average and range control limits for a specification midpoint value of 40.0.

SAMPLE		1	2	3	4	5	6	7
MEASUREMENT	1	40.4	41.0	41.4	40.2	40.1	40.2	40.4
	2	40.0	41.2	41.3	40.7	40.0	38.2	38.6
	3	40.1	39.4	40.0	41.1	40.5	39.8	40.0
	4	39.6	39.9	40.1	38.4	39.1	40.9	40.1
	5	39.3	40.5	39.9	39.5	40.8	39.5	38.8
TOTALS		199.4	202.0	202.7	199.9	200.5	198.6	197.9
$\overline{X}$		39.8	40.4	40.5	39.9	40.1	39.7	39.5
R		1.1	1.8	1.5	2.7	1.7	2.7	1.8

$\overline{\overline{X}} = 39.9 \quad \overline{R} = 1.9$

Average:

$$U_{CL} = SMP + (A_2 \times \overline{R}) = 40.0 + (.58 \times 1.9) = 41.1$$
$$L_{CL} = SMP - (A_2 \times \overline{R}) = 40.0 - (.58 \times 1.9) = 38.8$$

Range:

$$U_{CL} = D_4 \times \overline{R} = 2.11 \times 1.9 = 4.00$$
$$L_{CL} = D_3 \times \overline{R} = 0 \times 1.9 = 0.0$$

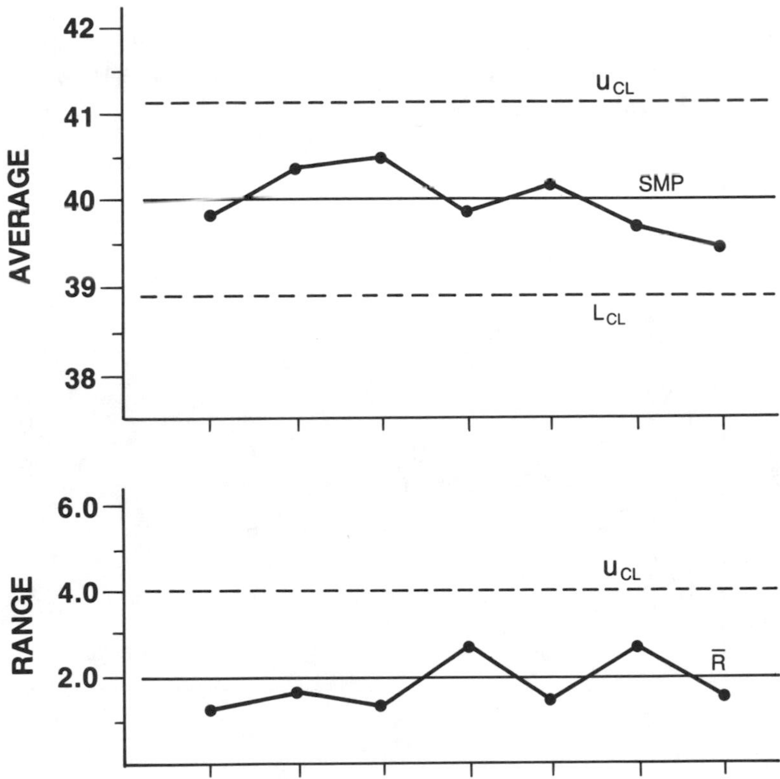

CONTROL CHART INTERPRETATION

If an event has an equal opportunity to occur, like getting a head or a tail on a single toss of a coin, the probabilities are one-half plus one-half. There is a 50% chance of getting a head, a 50% chance of getting a tail, and a 100% chance of getting either a head or a tail.

If an event is made up of a series of events, like getting two heads or two tails in a row, the probability of that event occurring is the product of the probabilities for a single event. Thus, the probability of getting two heads or two tails in a row is .5 times .5 or .25. Accordingly, the probability of getting three heads or three tails in a row is .5 times .5 times .5 or .125.

Applying these probabilities to the sampling distribution, the probability of a single sample average occurring above or below the mean is .5 or 50%. And the probability of getting two consecutive sample averages either above or below the mean is .5 times .5 or 25%.

This means that there is only a 25% chance of two consecutive sample averages occurring either above or below the mean, due to chance alone, unless the mean of the process has shifted (Figure 4.8).

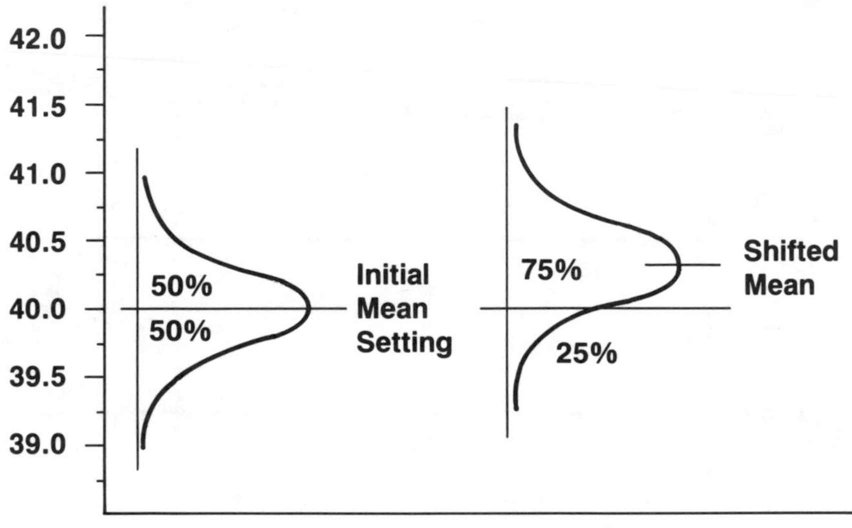

FIGURE 4.8

In practical terms, this means that if a greater precentage of sample averages occur either above the control chart midpoint, or below it, there is a good chance that the process mean has shifted.

By establishing the control chart limits at $\pm$ 3 standard deviations of the sampling distribution mean, the assumption is that corrective action measures will be taken when there is a 99.87% chance of an assignable cause process shift.

Turning this corrective action statement around (100 − 99.87), the decision to take action is when the element of chance error is only 0.13%; that is, when there is only a 0.13% chance that a sample average will fall outside the control limits due to chance alone.

Accordingly, other conditions which could result in a chance happening of only 0.13% can be determined by dividing the control chart into 1, 2, and 3 standard deviations above and below the midpoint (Figure 4.9).

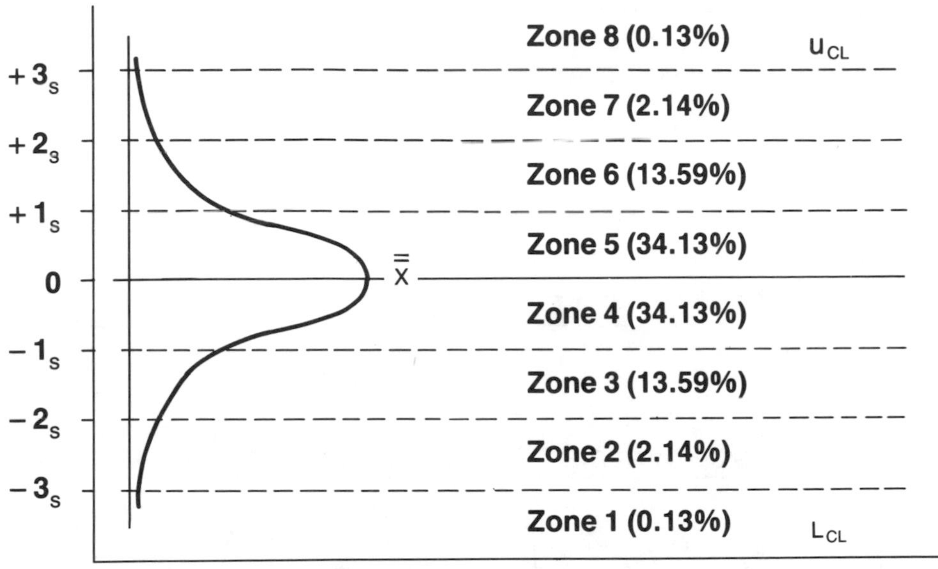

FIGURE 4.9

By multiplying the probability values associated with the various zones, the product probability rule, the required consecutive sample averages resulting in a chance happening of 0.13% are:

Above or Below the Mean: $(.5)^{10} = 0.09\%$
Zones 2 or 7: $(.0214)^2 = 0.04\%$
Zones 3 or 6: $(.1359)^3 = 0.25\%$
Zones 4 or 5: $(.3413)^6 = 0.15\%$

The first and last two terms of the binomial expansion can also be used to calculate non-consecutive sample averages. For example, the probability of 10 out of 11 sample averages falling on the same side of the midpoint is:

First Two Terms: $(.5)^{11} + 11(.5)^{10}(.5)$
Last Two Terms: $11(.5)(.5)^{10} + (.5)^{11}$

And since the probability of 10 out of 11 is the sum of the first two and last two terms, we have:

$$[.00048 + .00537] + [.00537 + .00048] = 1.17\%$$

Patterns of Nonrandom Variability

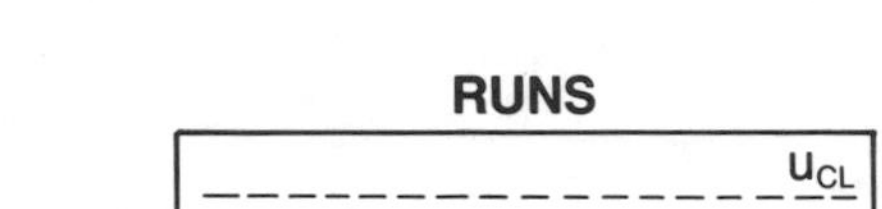

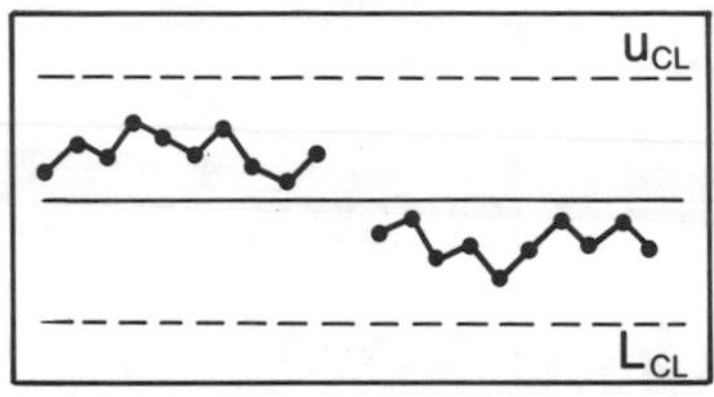

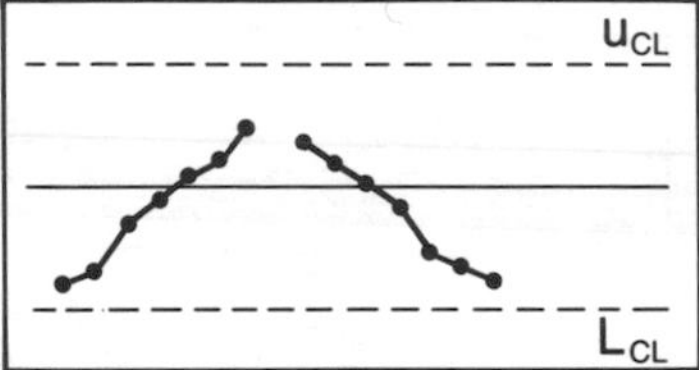

HUGGING CENTER LINE

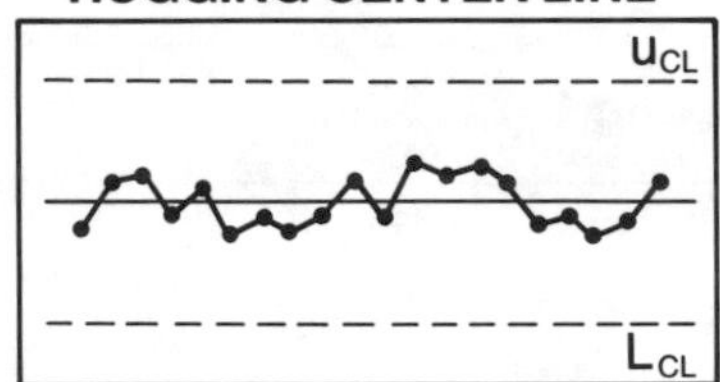

HUGGING CONTROL LIMITS

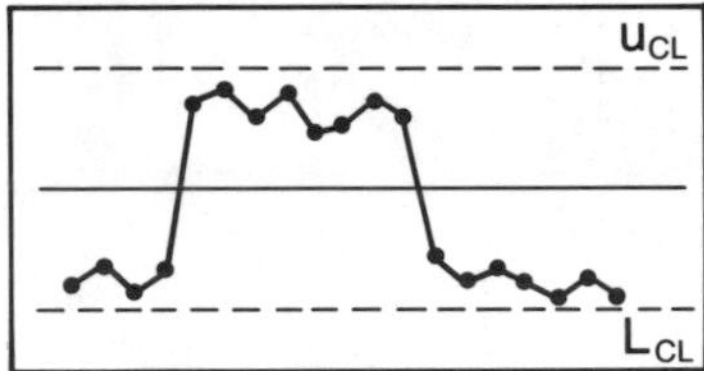

ZONE

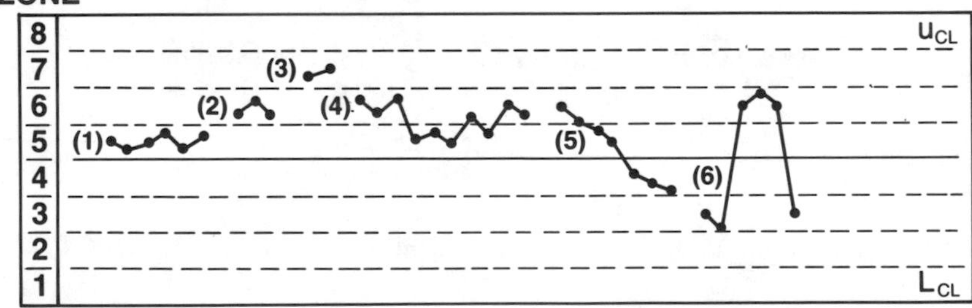

(1) 6 CONSECUTIVE POINTS IN ZONE 4 OR 5.

(2) 3 CONSECUTIVE POINTS IN ZONE 3 OR 6.

(3) 2 CONSECUTIVE POINTS IN ZONE 2 OR 7.

(4) 10 CONSECUTIVE POINTS ABOVE OR BELOW THE CENTER LINE.

(5) 7 CONSECUTIVE POINTS THAT RISE OR FALL.

(6) 6 CONSECUTIVE POINTS FALLING OUTSIDE ZONES 4 AND 5.

Variability and Dispersion

Since both process dispersion and process average may fluctuate over time, both the average and range control charts are needed to effectively monitor and control product quality (Figure 4.10).

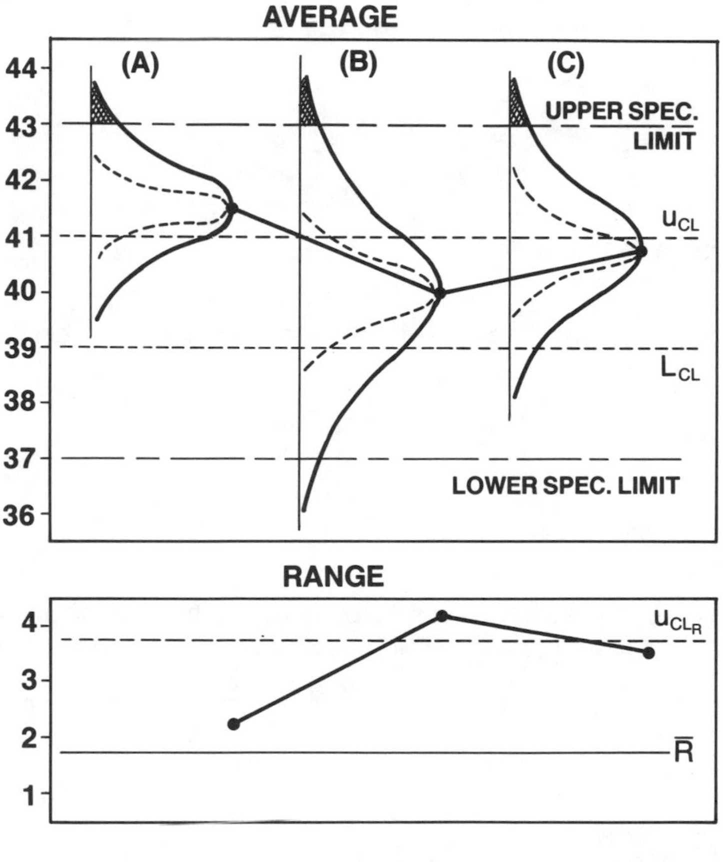

FIGURE 4.10

CONTROL CHARTS FOR INDIVIDUAL MEASUREMENTS

Although control charts for individual measurements are less satisfactory than for subgroup sample sizes of 5 or more, there are certain instances where only this type of chart can be used (e.g., to monitor the temperature of a plating tank). In this case, since tank temperature isn't likely to change appreciably in a short period of time, the average and range of 5 consecutive measurements would be meaningless.

Instead, individual temperature readings would be taken at periodic intervals and recorded on an individual measurement, or X, chart.

Control Limits

The formulas for computing the individual X chart control limits are essentially the same as for the $\overline{X}$ and R chart. In this case, however, since the sample size is 1, the constant multiplier for the average chart (A_2) is 2.66. Thus, the formulas for the average chart are:

$$U_{CL} = SMP + (A_2 \times \overline{R})$$
$$L_{CL} = SMP - (A_2 \times \overline{R})$$

where A_2 is equal to 2.66.

For the range chart, the range between the present and preceding measurement is used to establish the range value. Thus, for a sample of 2, the upper control limit formula is:

$$U_{CL} = \frac{3 \times \overline{R}}{d_2}$$

where d_2 is equal to 1.128.

Example 4.3
Given the following chemical measurements for a specification midpoint of 1.50% iron oxide, calculate average and range control limits.

SAMPLE	1	2	3	4	5	6	7
Value	1.49	1.51	1.48	1.50	1.49	1.53	1.50
Range		.02	.03	.02	.01	.04	.03

$\overline{X} = 1.50$ $\overline{R} = .025$

Average:

$$U_{CL} = SMP + (A_2 \times \overline{R}) = 1.50 + (2.66 \times .025) = 1.57$$
$$L_{CL} = SMP - (A_2 \times \overline{R}) = 1.50 - (2.66 \times .025) = 1.43$$

Range:

$$u_{CL} = \frac{3 \times \overline{R}}{d_2} = \frac{3 \times .025}{1.128} = .066$$

CONTROL CHARTS FOR ATTRIBUTES

While the $\overline{X}$, R, and X charts require actual numerical measurements, control charts for attributes require only a count of the number of observations in a particular category. For example, the number of items inspected or, in the case of defects, either the total number of defects in relation to the number of items inspected (defects per unit, per 100 units, etc.), or the total number of defects in relation to the total number of possible defects.

There are four basic types of attribute charts: the p chart for fraction or percent defective, np charts for the number of defectives, c charts for the number of defects, and u charts for the number of defects per unit.

Basic Theory

Like the $\overline{X}$ or average chart, the theory behind the attribute charts can be explained by visualizing a normal sampling distribution around a central or average point. While this is not exactly the case, since under normal operating circumstances the fraction and number defective charts (p and np) more closely follow the binomial distribution; while the charts for defects (c and u) follow the poisson distribution. In either case, the practical similarities are such that equal probabilities between the mean and the upper and lower control limits can be assumed.

Since a normal sampling distribution can be assumed for the attribute charts, upper and lower control limits can be established at ± 3 standard deviations from the mean (i.e., the average process percent defective for the p chart or, for the defect charts, the average number of defects). In this manner, as for the average chart, the area within the control limits would represent the area of normal random variation. The area outside the control limits would represent the area of assignable cause. Accordingly, if a sample value fell outside those limits, there would be a 99.73% chance that the process was producing greater or fewer defectives (or defects) than would be expected due to chance alone.

CONTROL CHART FOR PERCENT DEFECTIVE

When actual numerical measurements cannot be obtained, or when the economics of monitoring several characteristics is not feasible, the percent defective chart is often used to monitor quality.

While this is a more economical approach, since the p chart can monitor several characteristics at a time, it should be understood that the p chart is less sensitive to actual defect causes. Therefore, it is inferior to the $\overline{X}$ and R chart for assignable cause diagnosis.

If the p chart is properly designed, however, where the type and quantity of defects are recorded for each sample, it can provide useful defect analysis information. Thereby enhancing problem-solving efforts as well as for displaying trend information on the effectiveness of corrective action measures.

Sample Size

The first step in implementing a p chart is to determine the average percent defective currently produced by the process. This action is needed to determine the correct sample size for effectively discriminating between normal and assignable cause variation. For example, if the process average is 5% defective and a sample size of 10 is used, the best the sample could discriminate is that the process is either 0% defective (no defectives in the sample), or 10 or greater % defective (one or more defects in the sample). For this reason, a good rule of thumb is to use a sample size that is four times the quantity required to observe a single defect. Thus, a sample size of 80 would be the correct size for a process average of 5% defective.

Trial Control Limits

If the process percent defective is unknown, the usual procedure is to draw a random sample of at least 100 items from the process and compute the percent defective. This is accomplished by simply dividing the number of defectives observed in the sample by the number of samples inspected:

$$p = \frac{f}{n}$$

where p is the fraction defective, f is the number defective, and n is the sample size.

Using the calculated (assumed) process average to establish the ongoing sample size (four times the number to detect a single defective), the next step is to calculate the upper and lower trial control limits. This is accomplished by computing the sample standard deviation of the ongoing sample size and setting the control limits at ±3 standard deviations from the mean. For example, assuming a process percent defective value of 2%, which would result in a sample size of 200, the sample standard deviation would be:

$$S_p = \sqrt{\frac{pq}{n}} = \sqrt{\frac{(.02)(.98)}{200}} = .0098$$

where S_p is the standard deviation, p is the process fraction defective, q is the process fraction acceptable $(1 - p)$, and n is the sample size. Then, since the control limits are set at ±3 standard deviations from the mean, the upper and lower trial control limits are:

$$u_{CL} = .02 + 3(.0098) = .049 = 4.9\%$$

$$L_{CL} = .02 - 3(.0098) = -.009 = 0\%$$

The p chart, with trial control limits, would then be used to gather process information over a given period of time; usually one week to a month depending on the production volume and/or sampling frequency. At the end of the prescribed period, the average fraction defective $(\overline{p})$ for the combined samples would be used to compute new control limits.

Control Limits

The formulas for the upper and lower p chart control limits are essentially the same as for the trial control limits. The difference now being that $\overline{p}$ is used instead of p (i.e., the average fraction defective, as based on the combined periodic samples, is used in place of the single sample p value).

$$u_{CL} = \overline{p} + 3\sqrt{\frac{\overline{p}\,\overline{q}}{n}}$$

$$L_{CL} = \overline{p} - 3\sqrt{\frac{\overline{p}\,\overline{q}}{n}}$$

where $\overline{q}$ equals $1 - \overline{p}$ and n is the subgroup sample size.

Example 4.4

For a single sample of 100 products, a particular process is assumed to be 2% defective. Based on this information, a p chart with trial control limits is used to gather further process information. After inspecting a total of 12 200-piece samples and observing 74 total defects, compute the true process control limits.

$$\overline{p} = \frac{f}{n} = \frac{74}{2400} = .03$$

$$u_{CL} = \overline{p} + 3\sqrt{\frac{\overline{p}\,\overline{q}}{n}} = .03 + 3\sqrt{\frac{(.03)(.97)}{200}} = .066$$

$$= 6.6\%$$

$$L_{CL} = \overline{p} - 3\sqrt{\frac{\overline{p}\,\overline{q}}{n}} = .03 - 3\sqrt{\frac{(.03)(.97)}{200}} = -\ .006$$

$$= 0$$

As seen, even with a sample size of 200, the p chart is relatively insensitive in detecting significant process changes (Figure 4.11).

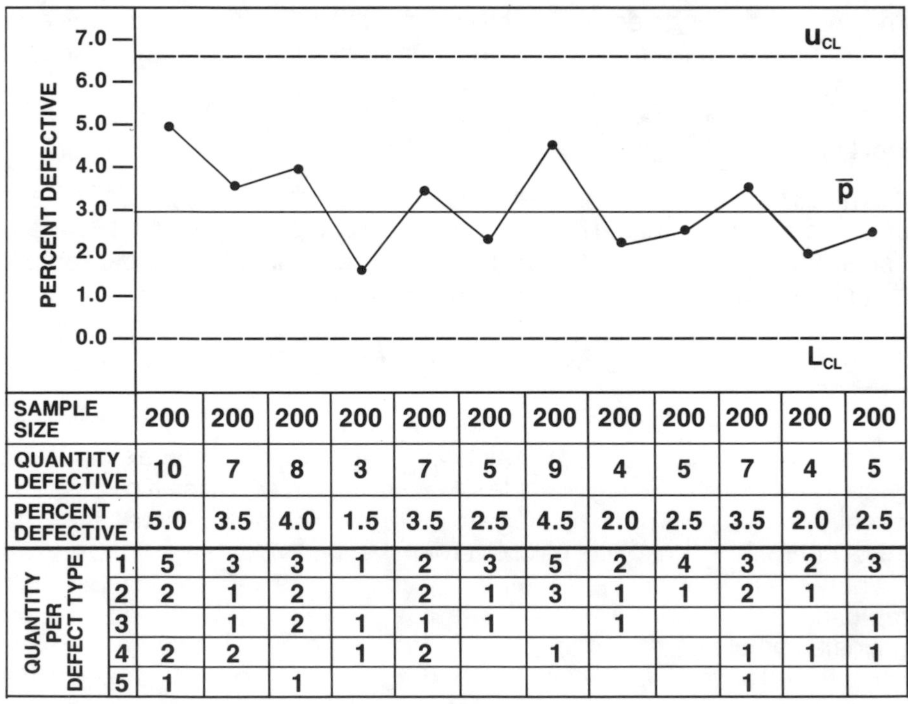

SAMPLE SIZE		200	200	200	200	200	200	200	200	200	200	200	200
QUANTITY DEFECTIVE		10	7	8	3	7	5	9	4	5	7	4	5
PERCENT DEFECTIVE		5.0	3.5	4.0	1.5	3.5	2.5	4.5	2.0	2.5	3.5	2.0	2.5
QUANTITY PER DEFECT TYPE	1	5	3	3	1	2	3	5	2	4	3	2	3
	2	2	1	2		2	1	3	1	1	2	1	
	3		1	2	1	1	1		1				1
	4	2	2		1	2		1			1	1	1
	5	1		1							1		

FIGURE 4.11

CONTROL CHART FOR NUMBER DEFECTIVE: np CHART

The major advantage of the np chart, as compared to the p chart, is that it eliminates calculating the percent defective value for each individual sample — i.e., if the sample size remains constant, the relationship between percent defective and number defective remains constant. Thus, the number defective can be plotted on the chart without converting to a percentage.

If the sample size remains constant, the standard deviation formula for percent defective ($sp = \sqrt{pq/n}$), can be converted to $snp = \sqrt{npq}$. Then, since the upper and lower control limits are now set at ± 3 standard deviations from the average number of defectives, the control limit formulas are simply:

$$U_{CL} = n\bar{p} + 3\sqrt{n\bar{p}\,\bar{q}}$$

$$L_{CL} = n\bar{p} - 3\sqrt{n\bar{p}\,\bar{q}}$$

Example 4.5
For a constant sample size of 200, drawn from a process with an average fraction defective of .03, compute upper and lower control limits for an np chart (Figure 4.12).

$$\begin{aligned} U_{CL} &= (200)(.03) + 3\sqrt{(200)(.03)(.97)} \\ &= 6 + 7.2 = 13.2 \end{aligned}$$

$$\begin{aligned} L_{CL} &= (200)(.03) - 3\sqrt{(200)(.03)(.97)} \\ &= 6 - 7.2 = 0 \end{aligned}$$

CONTROL CHART FOR DEFECTS: c CHART

Control charts for defects, the c chart, can be used to monitor either the number of defects for a single sample unit or the total number of defects for a sample group. For sample groups, however, the sample size must remain constant. If not, the control chart center line ($\bar{c}$), which is the average total number of defects, as well as the upper and lower control limits, must be calculated for each individual sample size. In either case, whether single or multiple samples, the c chart is most applicable when the opportunity for a defect to occur is large and the probability of detecting a defect is small (i.e., the poisson distribution).

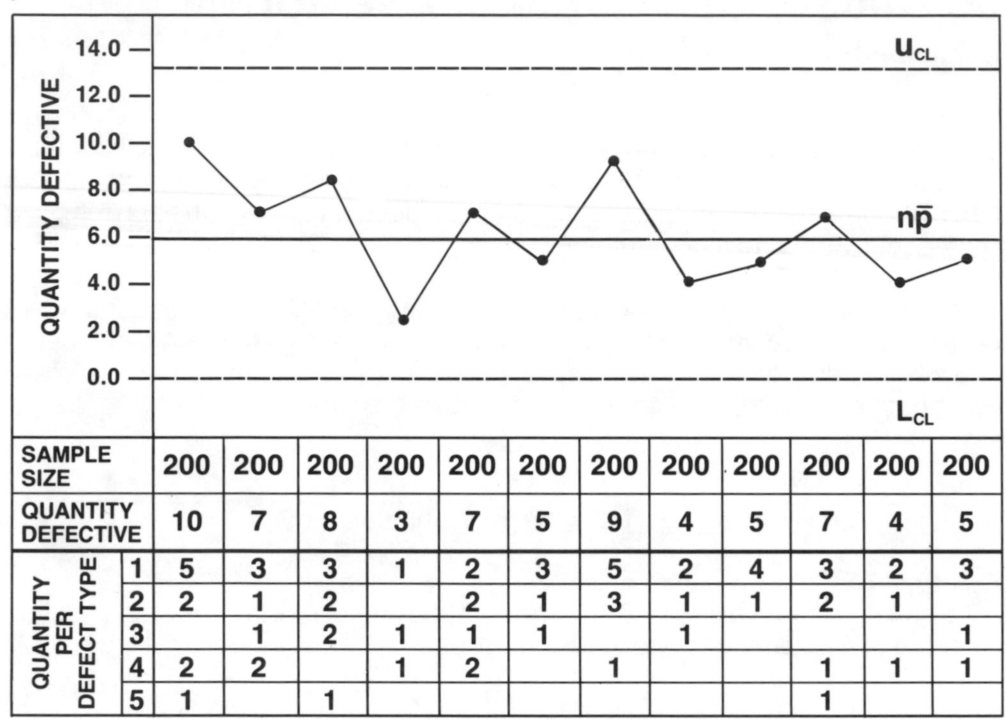

SAMPLE SIZE		200	200	200	200	200	200	200	200	200	200	200	200
QUANTITY DEFECTIVE		10	7	8	3	7	5	9	4	5	7	4	5
QUANTITY PER DEFECT TYPE	1	5	3	3	1	2	3	5	2	4	3	2	3
	2	2	1	2		2	1	3	1	1	2	1	
	3		1	2	1	1	1		1				1
	4	2	2		1	2		1			1	1	1
	5	1		1							1		

FIGURE 4.12

The standard deviation of the sampling distribution of defects is simply the square root of the average expected total defects or $s_{\bar{c}} = \sqrt{\bar{c}}$. Thus, the upper and lower control limit formulas are:

$$u_{CL} = \bar{c} + 3\sqrt{\bar{c}}$$

$$L_{CL} = \bar{c} - 3\sqrt{\bar{c}}$$

Example 4.6

For a constant sample size of 10 and an average total number of defects of 5, compute the upper and lower control limits for a c chart (Figure 4.13).

$$u_{CL} = 5 + 3\sqrt{5} = 11.7 = 11$$

$$L_{CL} = 5 - 3\sqrt{5} = -1.7 = 0$$

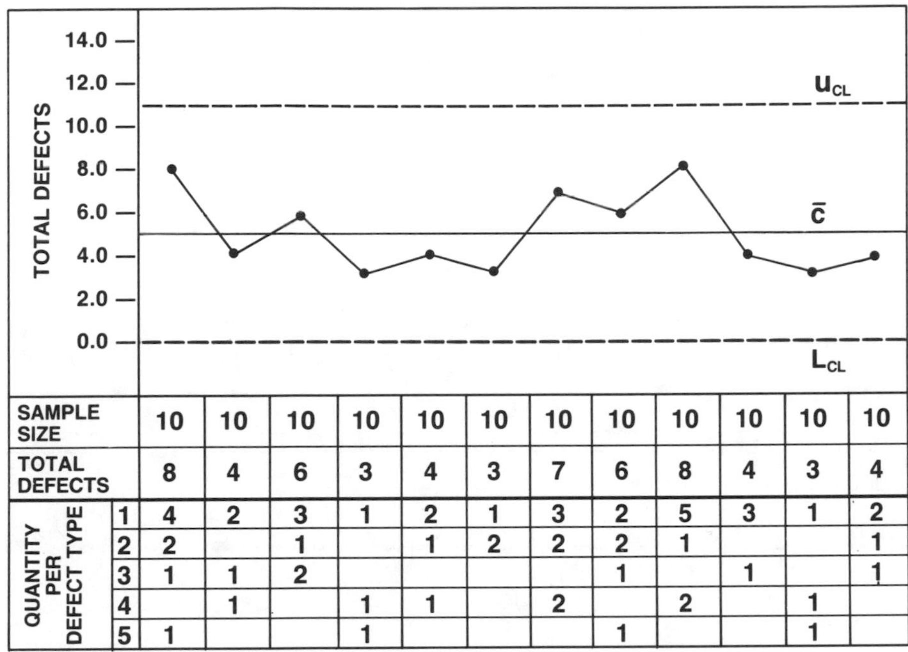

		1	2	3	4	5	6	7	8	9	10	11	12
SAMPLE SIZE		10	10	10	10	10	10	10	10	10	10	10	10
TOTAL DEFECTS		8	4	6	3	4	3	7	6	8	4	3	4
QUANTITY PER DEFECT TYPE	1	4	2	3	1	2	1	3	2	5	3	1	2
	2	2		1		1	2	2	2	1			1
	3	1	1	2					1		1		1
	4		1		1	1		2		2		1	
	5	1			1				1			1	

FIGURE 4.13

CONTROL CHART FOR AVERAGE DEFECTS: u CHART

The u chart is simply a modified c chart where the average number of defects are plotted instead of the total number of defects. The approach provides a constant control chart center line (u) regardless of the size of the sample. For example, for a sample size of 2 having 2 defects each, the total number of defects is 4 while the average number of defects is 2. Accordingly, a sample size of 5, each having 2 defects, results in a total of 10 defects while maintaining an average of 2 defects.

The standard deviation of the defect distribution, as used in computing the c chart control limits, is $\sqrt{\bar{c}}$. For the u chart, however, since u equals $\bar{c}$, or c divided by n, the upper and lower control limit formulas are:

$$u_{CL} = u + 3\sqrt{\frac{u}{n}}$$

$$L_{CL} = u - 3\sqrt{\frac{u}{n}}$$

Example 4.7
Based on the inspection of 60 products, 240 defects were observed. For a control chart sample of 5, compute u and upper and lower control limits (Figure 4.14).

$$u = \frac{c}{n} = \frac{240}{60} = 4.0$$

$$u_{CL} = 4.0 + 3\sqrt{\frac{4.0}{5}} = 6.68$$

$$L_{CL} = 4.0 - 3\sqrt{\frac{4.0}{5}} = 1.31$$

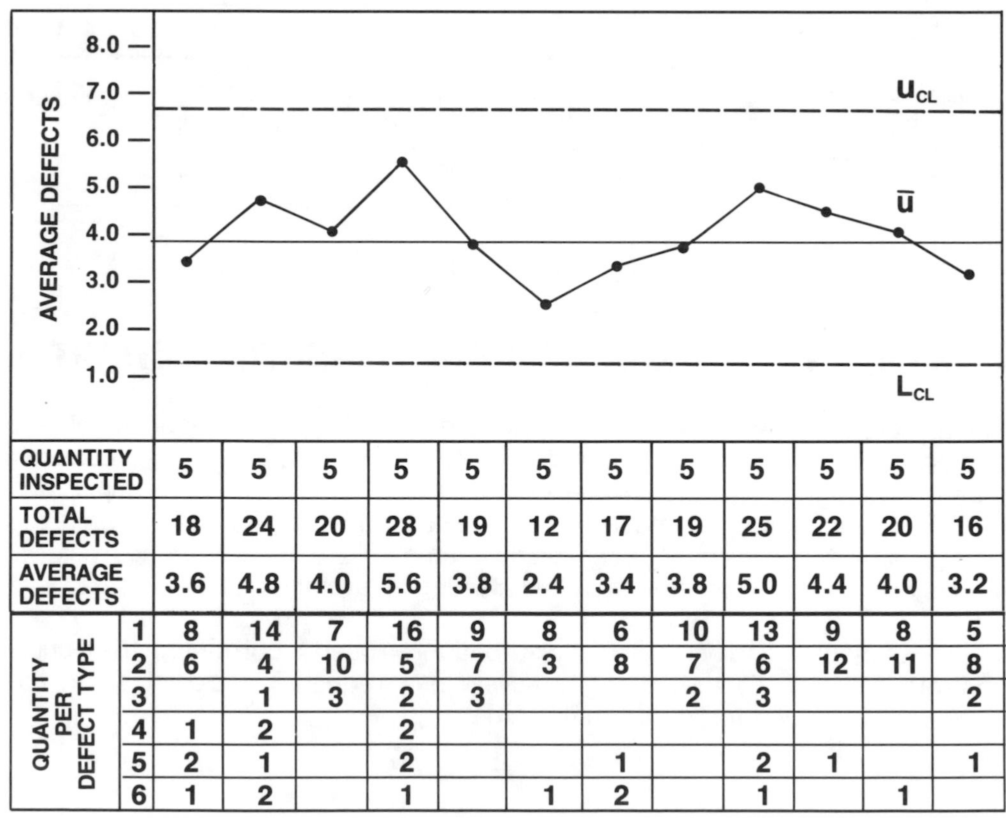

QUANTITY INSPECTED	5	5	5	5	5	5	5	5	5	5	5	5
TOTAL DEFECTS	18	24	20	28	19	12	17	19	25	22	20	16
AVERAGE DEFECTS	3.6	4.8	4.0	5.6	3.8	2.4	3.4	3.8	5.0	4.4	4.0	3.2
QUANTITY PER DEFECT TYPE 1	8	14	7	16	9	8	6	10	13	9	8	5
2	6	4	10	5	7	3	8	7	6	12	11	8
3		1	3	2	3			2	3			2
4	1	2		2								
5	2	1		2			1		2	1		1
6	1	2		1		1	2		1		1	

FIGURE 4.14

CONTROL CHART FOR DEMERITS: D CHART

A useful adaptation of either the control chart for total defects (c chart) or control chart for average defects (u chart), is the control chart for demerits.

This is accomplished by classifying possible defects into categories such as critical, major, and minor. Relative numerical weights would then be assigned to each category and, based on inspection information or past history, a control chart constructed to monitor either total demerits (c chart) or average demerits (u chart).

Defect Classification	Relative Weight
Critical	5
Major	3
Minor	1

A single product having 1 critical defect, 2 major defects, and 3 minor defects, would have a total demerit value of 14.

Example 4.8
Based on a sample of 10 products, 8 critical, 22 major, and 54 minor defects were observed. Using a numerical weighting of 5 for critical defects, 3 for major defects, and 1 for minor defects, calculate the upper and lower control limits for a total demerits control chart.

$$\text{Total demerits (D)} = (8 \times 5) + (22 \times 3) + (54 \times 1) = 160$$

$$\overline{D} = \frac{D}{n} = \frac{160}{10} = 16$$

$$u_{CL} = \overline{D} + 3\sqrt{\overline{D}} = 16 + 3\sqrt{16} = 28$$

$$L_{CL} = \overline{D} - 3\sqrt{\overline{D}} = 16 - 3\sqrt{16} = 4$$

Exercise Worksheets

Exercise 4.1

AVERAGE AND RANGE CONTROL CHART

An $\overline{X}$ and R chart is to be used to monitor a critical dimensional characteristic of .040 ± .003 inch. Use the listed sample measurements to construct a control chart. When completed, plot the sample average and range values on the chart and answer the questions at the bottom of page 69.

SAMPLE MEASUREMENTS

	.040	**.039**	**.040**	**.035**	**.039**
	.038	**.040**	**.042**	**.037**	**.041**
	.039	**.040**	**.040**	**.041**	**.042**
	.040	**.041**	**.041**	**.042**	**.038**
	.038	**.043**	**.042**	**.039**	**.037**
$\overline{X}$ =	_____	_____	_____	_____	_____
R =	_____	_____	_____	_____	_____
	.039	.040	.045	.040	.040
	.042	.043	.041	.041	.037
	.040	.041	.040	.039	.038
	.043	.040	.039	.038	.040
	.038	.042	.040	.039	.041
$\overline{X}$ =	_____	_____	_____	_____	_____
R =	_____	_____	_____	_____	_____

$\overline{R} = [\quad\quad]$

$U_{CL_{\overline{X}}} = SMP + A_2 \times \overline{R} = [\quad\quad]$

$U_{CL_R} = D_4 \times \overline{R} = [\quad\quad]$

$L_{CL_{\overline{X}}} = SMP - A_2 \times \overline{R} = [\quad\quad]$

Exercise 4.1 (continued)

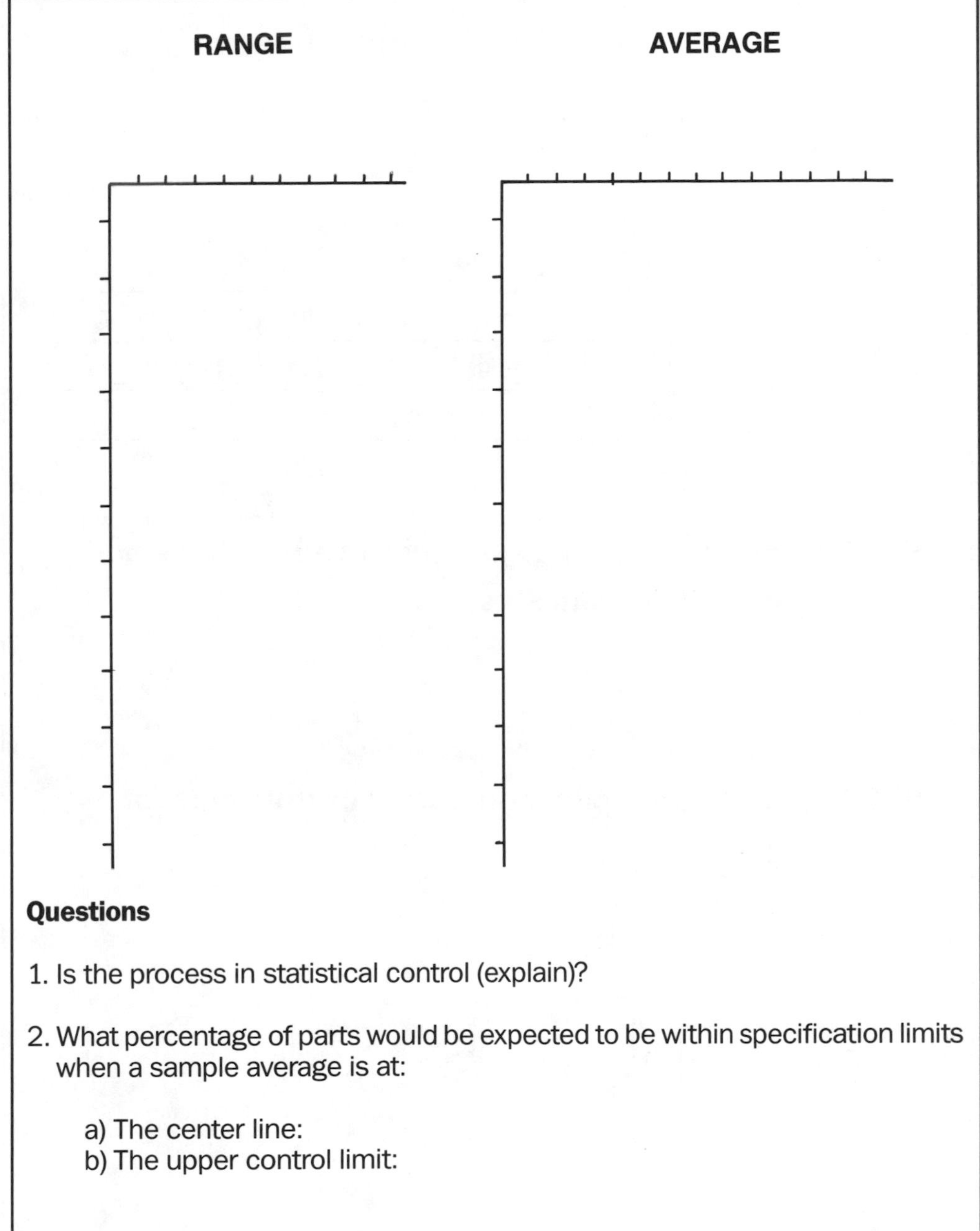

Questions

1. Is the process in statistical control (explain)?

2. What percentage of parts would be expected to be within specification limits when a sample average is at:

   a) The center line:
   b) The upper control limit:

Exercise 4.2

INDIVIDUAL (X) CHART

Given the following daily solder temperature readings, for a wave solder machine having a specification midpoint of 500°F with a tolerance of ±25°F, construct a control chart for individual measurements.

SAMPLE	1	2	3	4	5	6	7
Value	507	502	495	501	494	498	503
Range							

STEP 1: Enter sample range values in above table and compute $\overline{X}$ and $\overline{R}$ values.

$\overline{X} = [\quad\quad]$ $\overline{R} = [\quad\quad]$

STEP 2: Compute upper and lower control limits for average.

$$U_{CL} = SMP + (A_2 \times \overline{R})$$

$$= [\quad\quad] + ([\quad\quad] \times [\quad\quad]) = [\quad\quad]$$

Exercise 4.2 (continued)

$$L_{CL} = SMP - (A_2 \times \bar{R})$$

$$= [\quad] - ([\quad] \times [\quad]) = [\quad]$$

STEP 3: Compute upper control limit for range.

$$U_{CL} = \frac{3 \times \bar{R}}{d_2} = \frac{3 \times [\quad]}{[\quad]} = [\quad]$$

STEP 4: Construct control chart and plot sample values.

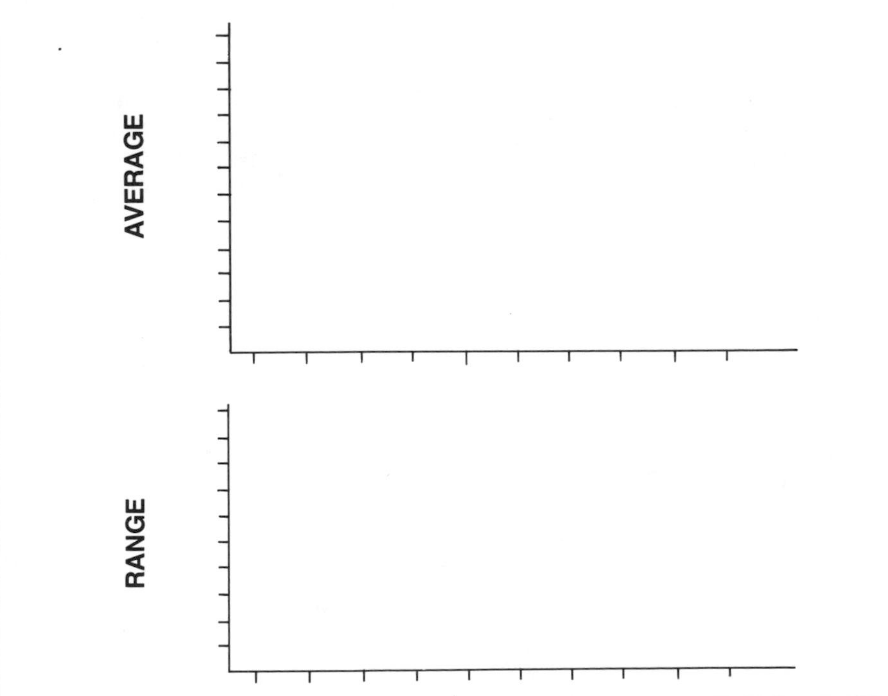

Exercise 4.3

PERCENT DEFECTIVE CONTROL CHARTS

End-line inspection data indicate that a particular process is producing an average rate of 5% defective products. To guard against short-term high percentages, and to isolate assignable causes, a p chart is to be used to monitor the process on an ongoing basis.

Given the above process average of 5% defective, determine the appropriate p chart sample size and calculate the upper and lower control limits.

Sample size = [ ]

Note: The sample size should be at least four times the quantity needed to detect a single defect.

Exercise 4.3 (continued)

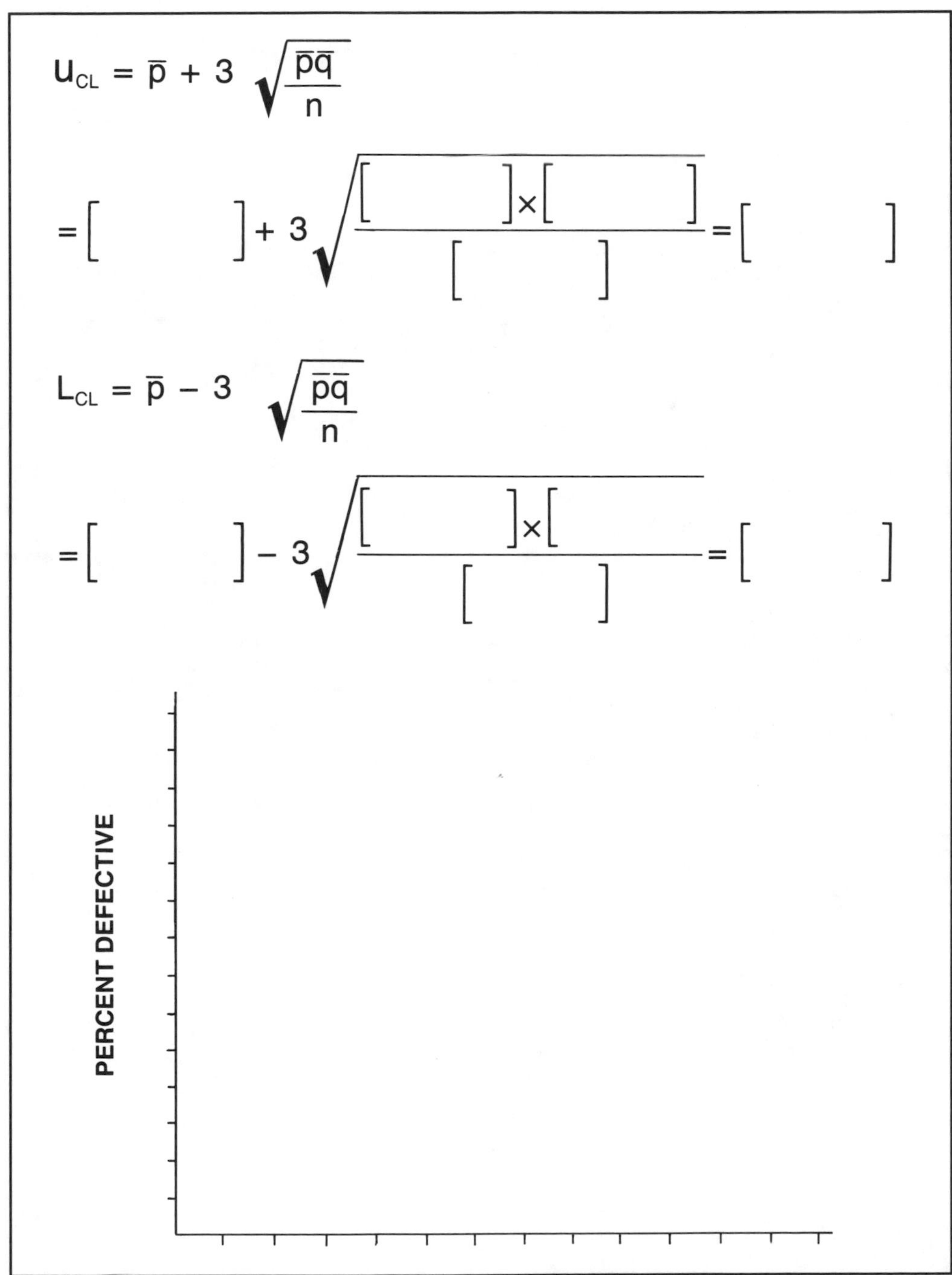

Exercise 4.4

NUMBER DEFECTIVE CONTROL CHARTS

Given the same process as used for the p chart (i.e., 5% defective and a sample size of 80), construct an np chart to eliminate the need to compute individual sample percentages.

$$U_{CL} = n\bar{p} + 3\sqrt{n\bar{p}\bar{q}}$$

$$= [\quad] \times [\quad] + 3\sqrt{[\quad] \times [\quad] \times [\quad]}$$

$$= [\quad]$$

$$L_{CL} = n\bar{p} - 3\sqrt{n\bar{p}\bar{q}}$$

$$= [\quad] \times [\quad] - 3\sqrt{[\quad] \times [\quad] \times [\quad]}$$

$$= [\quad]$$

NUMBER DEFECTIVE

Exercise 4.5

CONTROL CHART FOR DEFECTS

Over a period of 6 weeks, a total of 25 defects were recorded for the final inspection of 15 finished products. Using these data, construct a defects per unit control chart.

$$\bar{C} = \frac{c}{n} = \frac{[\quad]}{[\quad]} = [\quad]$$

$$U_{CL} = \bar{c} + 3\sqrt{\bar{c}} = [\quad] + 3\sqrt{[\quad]} = [\quad]$$

$$L_{CL} = \bar{c} - 3\sqrt{\bar{c}} = [\quad] - 3\sqrt{[\quad]} = [\quad]$$

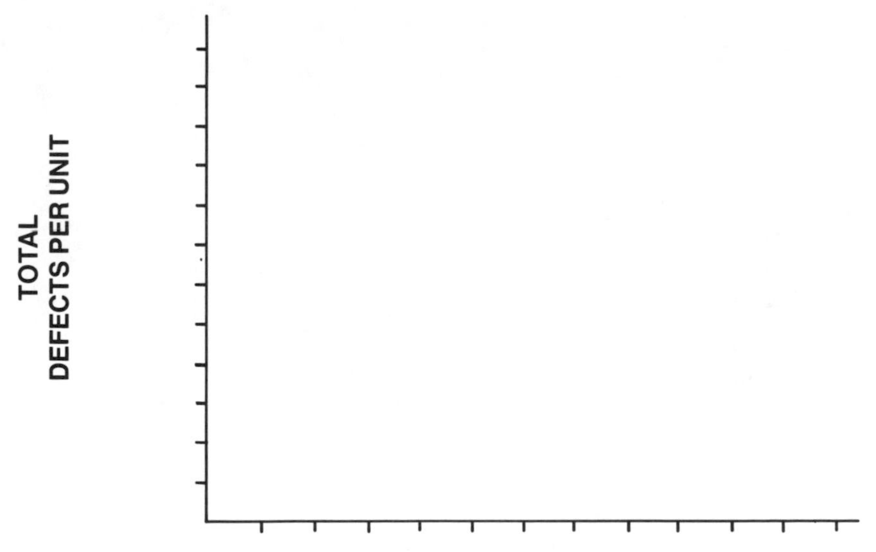

Exercise 4.6

CONTROL CHART FOR AVERAGE DEFECTS

One hundred percent inspection of printed circuit boards yields an average of 5 defects per board. Using these data, construct a control chart for the average number of defects for a sample grouping of 10 boards.

$$u = \frac{\bar{c}}{n} = \frac{[\quad\quad]}{[\quad\quad]} = [\quad\quad]$$

$$U_{CL} = u + 3\sqrt{\frac{u}{n}} = [\quad\quad] + 3\sqrt{\frac{[\quad\quad]}{[\quad\quad]}} = [\quad\quad]$$

$$L_{CL} = u - 3\sqrt{\frac{u}{n}} = [\quad\quad] - 3\sqrt{\frac{[\quad\quad]}{[\quad\quad]}} = [\quad\quad]$$

AVERAGE DEFECTS PER UNIT

Exercise 4.7

CONTROL CHART FOR DEMERITS

Inspection of 10 final products resulted in the observation of 3 critical defects, 13 major defects, and 25 minor defects. Given a weighting factor of 10 for critical defects, 5 for major defects, and 1 for minor defects, construct a control chart for demerits.

$$\bar{D} = \frac{D}{n} = \frac{[\quad\quad]}{[\quad\quad]} = [\quad\quad]$$

$$U_{CL} = \bar{D} + 3\sqrt{\bar{D}} = [\quad\quad] + 3\sqrt{[\quad\quad]} = [\quad\quad]$$

$$L_{CL} = \bar{D} - 3\sqrt{\bar{D}} = [\quad\quad] - 3\sqrt{[\quad\quad]} = [\quad\quad]$$

SECTION 5

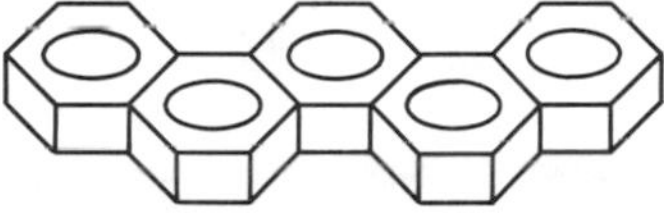

DATA GATHERING, PRESENTATION, AND BASIC PROBLEM SOLVING

PURPOSE OF DATA

Business activities typically involve working with a great deal of numerical data, but reams of raw data can be confusing rather than enlightening. A means of organizing and displaying data so that they can be properly interpreted and understood is needed. This is especially true when presenting situations that require management's attention. If we can remove the clutter and, instead, zero in on the net effects or root causes of what we're trying to convey, then the points we want to make are more likely to be understood and acted upon.

There are several useful forms for organizing and displaying data. These include frequency distributions, histograms, frequency polygons, trend charts, and many more. This section will review the basic concepts and variations of these data summary techniques and then discuss their use in root-cause problem analysis.

FREQUENCY DISTRIBUTIONS

A frequency distribution is simply a set of observations that show how often something occurs. For example, 10 shipping documents at a distribution center are reviewed to assess the number of on-time versus late shipments. Of these 10, the number (or frequency) of on-time shipments is 7 and the frequency of late shipments is 3. These data can be arranged in the following form where f denotes the frequency (Figure 5.1):

	f
On-time	7
Late	3
Total	10

FIGURE 5.1

This arrangement of the data is a frequency distribution. It is an arrangement of the data that shows how often on-time and late shipments occur.

Another example of a frequency distribution is the number of days between the placement of a purchase order and receipt of the material. For this example, consider the data in Figure 5.2.

Procurement Cycle in Days									
66	47	74	57	49	47	24	57	29	64
36	50	27	57	54	31	48	19	43	47
45	48	68	20	38	31	39	56	39	53
48	36	34	45	42	40	30	40	37	37
40	30	70	53	54	55	69	51	46	60
64	37	55	47	41	16	47	30	40	79
61	66	41	50	23	38	60	28	32	52
64	53	23	29	43	40	56	43	68	29
42	50	55	40	31	50	38	51	37	58
46	52	50	56	49	46	61	48	27	56

FIGURE 5.2

Now, by counting the number of times the same days occur, the data can be arranged and presented in a more meaningful form, as shown in Figure 5.3.

Data presented in this form of frequency distribution are more easily interpreted than the simple listing of raw numbers. However, if the data are widespread and discontinuous, as in the previous example, this approach may fail to readily convey subtle aspects of the distribution. Thus, a more convenient approach is to group the data within class intervals.

ESTABLISHING CLASS INTERVALS

When dealing with large sets of diverse data, subtle aspects of a frequency distribution often can go undetected unless the data are grouped within class intervals, usually 10 to 15 depending upon the range of the data. Class intervals are arbitrary divisions of the data into subgroups.

Frequency Distribution of Days									
Days	f	Days	f	Days	f	Days	f	Days	f
80	-	67	-	54	2	41	2	28	1
79	1	66	2	53	3	40	6	27	2
78	-	65	-	52	2	39	2	26	-
77	-	64	3	51	2	38	3	25	-
76	-	63	-	50	5	37	4	24	1
75	-	62	-	49	2	36	2	23	2
74	1	61	2	48	4	35	-	22	-
73	-	60	2	47	5	34	1	21	-
72	-	59	-	46	3	33	-	20	1
71	-	58	1	45	2	32	1	19	1
70	1	57	3	44	-	31	3	18	-
69	1	56	4	43	3	30	3	17	-
68	2	55	3	42	2	29	3	16	1

FIGURE 5.3

For example, consider the previous frequency distribution of days from purchase order release to receipt of material. In this case, the data range from 16 to 79 days. Now, if we arbitrarily define five-day interval groupings, starting with the lowest value, we would assign all scores within the range of 15 to 19 to the first group, 20 to 24 to the second group, and so forth to the last group of 75 to 79. Such groupings of data, as shown in Figure 5.4, are displayed by entering a tally mark for each score opposite the class interval in which it falls. The tally marks are then counted to obtain the number of cases or frequency within each interval.

This arrangement of the data is just as much a frequency distribution as in the previous example. The only difference is that the class interval is 5 instead of 1. Thus, a frequency distribution is any arrangement of data which shows either the frequency of occurrence of individual values, or groups of values falling within arbitrarily defined class intervals.

Interval (days)	Tally	Frequency
75 - 79	I	1
70 - 74	II	2
65 - 69	IIII	5
60 - 64	IIII II	7
55 - 59	IIII IIII I	11
50 - 54	IIII IIII IIII	14
45 - 49	IIII IIII IIII I	16
40 - 44	IIII IIII III	13
35 - 39	IIII IIII I	11
30 - 34	IIII III	8
25 - 29	IIII I	6
20 - 24	IIII	4
15 - 19	II	2

FIGURE 5.4

The grouping of data into class intervals, however, may result in a loss of information pertaining to individual scores (i.e., scores may differ from one another within a limited range and still be grouped within the same interval). To minimize this effect, it is often helpful to select class intervals such that the total range of the data are covered with 10 to 15 intervals (keeping in mind that smaller sets of data may require fewer intervals than large sets). In general, the best criteria to follow is to select the number of class intervals (the number of cells) which best describe the distribution's true nature.

CLASS INTERVAL LIMITS AND MIDPOINTS

In the previous tally mark distribution, the apparent width of each interval is 4 (i.e., $19 - 15 = 4$ for the first interval, $24 - 20 = 4$ for the second interval, and so forth for the remaining intervals). The actual width for each interval, however, and to ensure that the interval midpoint is always a whole number, is defined by adding and subtracting one-half of the measurement value to the highest and lowest score of each interval. In this manner, we have 19.5 − 14.5 = 5 for the first interval, 24.5 − 19.5 = 5 for the second interval, and so on.

This distinction between apparent limits and real limits guarantees that the midpoint of each interval will be a whole number (i.e., in our example, the interval 14.5 to 19.5 is 5 units in width with a midpoint of 17). In taking this approach, however, we assume that the midpoint of each interval is the best single value to represent all values within the interval. While this may not always be the case, the class interval frequency distribution still provides the best method for graphically presenting large sets of data.

HISTOGRAMS

A histogram is probably the easiest way to present data graphically. It is constructed by scaling the class interval along the horizontal (X) axis of the graph and the frequency of occurrence along the vertical (Y) axis. To illustrate this construction, consider the data presented in the previous tally mark frequency distribution.

First, horizontal and vertical lines are drawn and scaled to appropriately represent the number of days and the frequency of occurrence. For these data, although the scale is arbitrary, if we allow 3/8 inch for each class interval and 3/16 inch for each unit of frequency, we obtain a graph approximately 5 inches wide by 3 inches high. The midpoints of each interval are then noted along the horizontal baseline and the frequency scaled along the vertical axis. Then, for each class interval, the corresponding frequency is plotted by drawing a horizontal line across the full length of the interval. The bar is then completed by joining the ends of these lines to the corresponding ends of the intervals along the horizontal axis.

When these actions are completed, the horizontal and vertical axis are appropriately labeled and, of particular importance, a concise statement given of what the chart illustrates (Figure 5.5).

FREQUENCY POLYGON

A frequency polygon is another common method of graphically portraying data. It is often used to indicate the general shape of data and for comparing its shape to the shape of hypothetical distributions. It also is quite useful in estimating whether data tend to be normally distributed about the midpoint, or skewed. Skewed means that more scores are at one end of the distribution than the other.

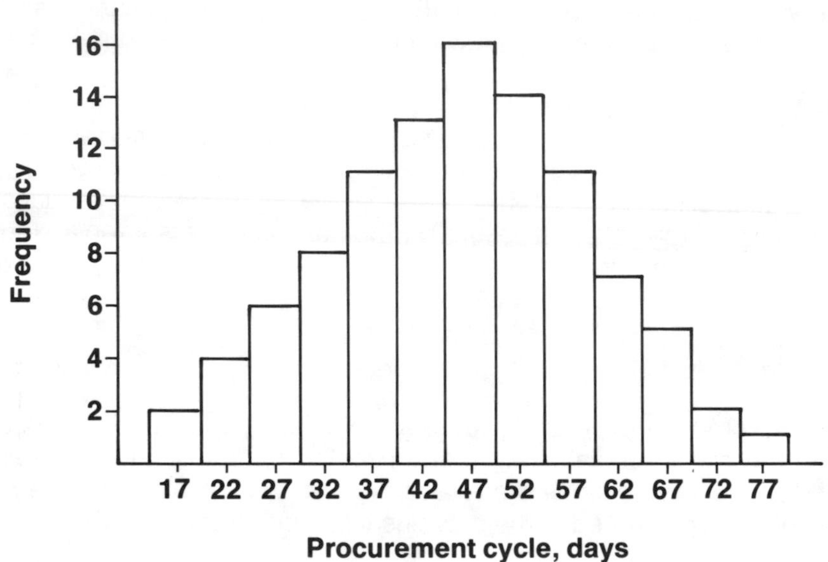

FIGURE 5.5

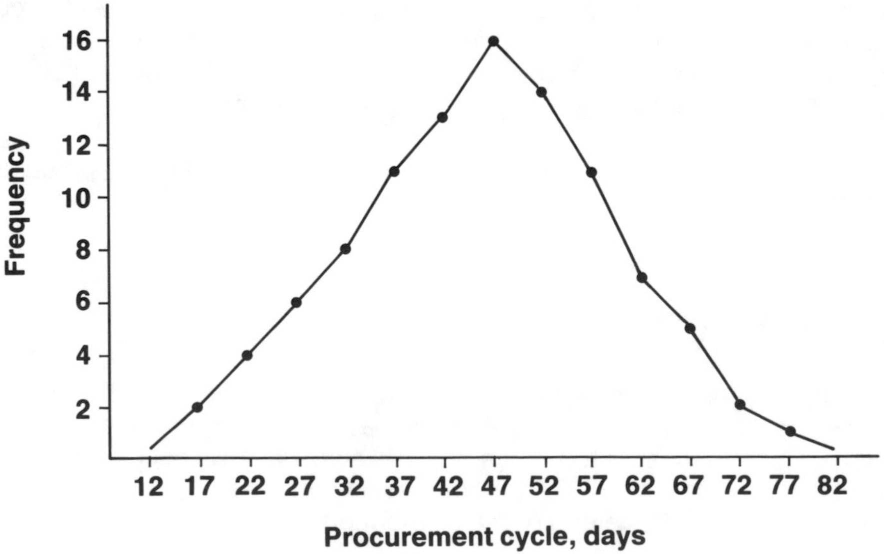

FIGURE 5.6

By drawing a frequency polygon, one can also determine if the distribution is flat or peaked. The flatness or peakedness of the data, commonly called kurtosis, is referred to as leptokurtic if it is peaked, mesokurtic if it is normal, and platykurtic if it is flat.

The basic difference between a frequency polygon and a histogram is the assumption that, for the frequency polygon, all occurrences are concentrated at the midpoint of the interval. For the histogram, the assumption is that they are evenly distributed across the range of each interval. Thus, instead of drawing a horizontal line the full length of the interval, a frequency polygon is constructed by placing a dot above the interval midpoint at a height proportional to the frequency of occurrence or percent of total ocurrences. The dots are then connected by drawing a solid line from one to the other. For the frequency polygon, it is also customary to show an additional interval at each end of the horizontal scale.

The frequency polygon shown in Figure 5.6 is based on the same data used to construct the procurement cycle histogram.

Observe that the frequency polygon in Figure 5.3 does not have a smooth continuous curve. This is because the dots are not only joined by straight lines, but because there are too few occurrences in each interval. This situation becomes less obvious as the range of the intervals becomes smaller and, at the same time, the total number of occurrences becomes larger. In the extreme situation, with indefinitely small intervals and an indefinitely large number of occurrences, the frequency polygon would take the shape of a continuous frequency distribution.

While the above frequency polygon tends to be somewhat symmetrical about the midpoint, other forms of data could result in frequency polygons that are skewed to the left or right, or exhibit varying degrees of kurtosis (Figures 5.7, 5.8, 5.9).

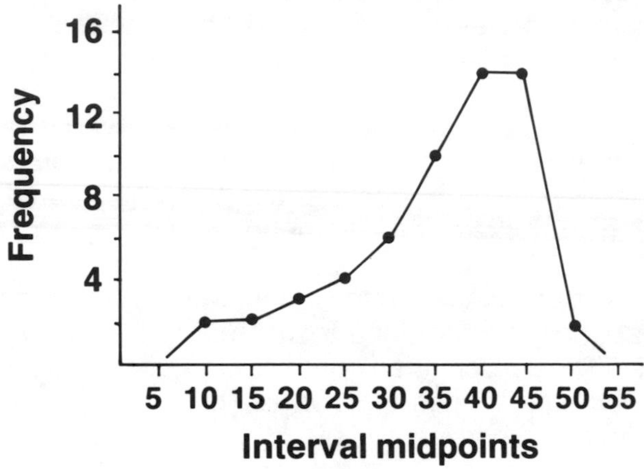

FIGURE 5.7

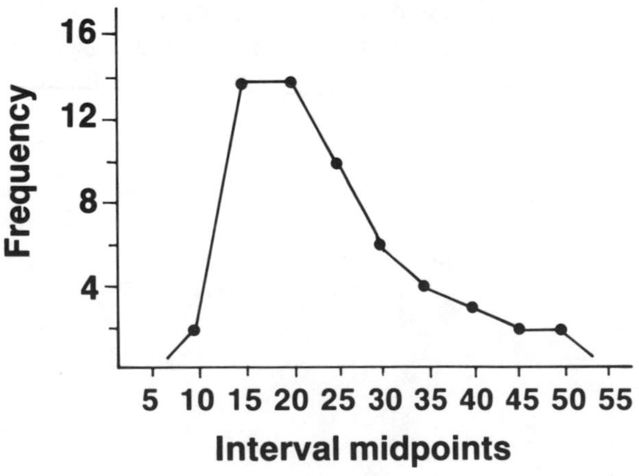

FIGURE 5.8

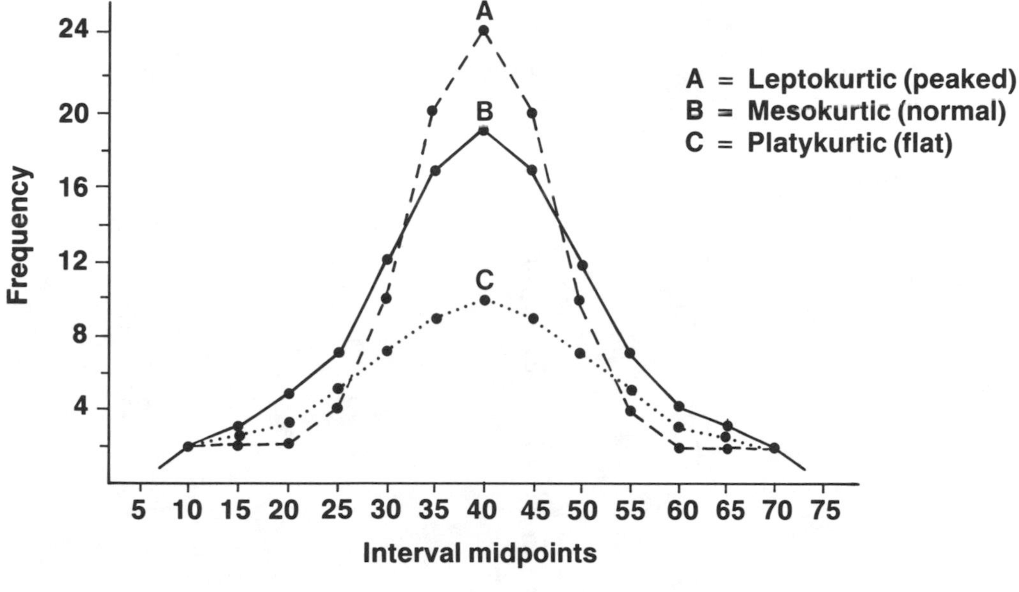

FIGURE 5.9

CUMULATIVE FREQUENCY POLYGONS

While it is always important to know the shape of data, situations may arise where it is equally important to graphically display the number or percentage of events falling above or below particular values. This is achieved by drawing a cumulative frequency polygon. Cumulative frequency polygons differ from standard frequency polygons in two respects: First, instead of plotting points corresponding to frequencies, they are plotted corresponding to cumulative frequencies. Second, to graphically represent the number of occurrences falling above or below particular values, the points are plotted above the upper interval boundary instead of above the interval midpoint. To illustrate this procedure, consider the data shown in Figure 5.10 representing the number of hours to complete customer service actions on a particular type of office machine.

To plot a cumulative frequency distribution for these data, we would plot the frequency 4 above the upper interval limit of 5, the cumulative frequency 19 above 10, and so forth for the remaining cumulative frequency values.

Interval (hours)	Interval midpoint	Freq.	Cumulative frequency	Cumulative percentage
51 - 55	53	1	210	100.0
46 - 50	48	3	209	99.5
41 - 45	43	5	206	98.0
36 - 40	38	7	201	95.7
31 - 35	33	13	194	92.3
26 - 30	28	19	181	86.1
21 - 25	23	31	162	77.1
16 - 20	18	72	131	62.3
11 - 15	13	40	59	28.0
6 - 10	8	15	19	9.0
1 - 5	3	4	4	1.9
Total		210		

FIGURE 5.10

By converting the cumulative frequencies to cumulative percentage frequencies, accomplished by dividing each cumulative frequency by the total frequency value of 210, we can also construct a cumulative percentage polygon. The advantage of this type of polygon is that the percentage of occurrences, for any specified period, can be read directly from the graph (i.e., approximately 2% of customer service actions are completed within five hours; 77% are completed within 25 hours; 23% of all customer service actions take greater than 26 hours to complete; all customer service actions are completed within 55 hours (Figure 5.11).

TREND CHARTS

Histograms and polygons are commonly used to graphically portray the frequency of occurrence of particular data values associated with a particular time period. When it is important to display particular categories over several time periods, however, trend or line charts are the appropriate method. These types of displays, for example, would include such categories as the variation in central tendency from one dis-

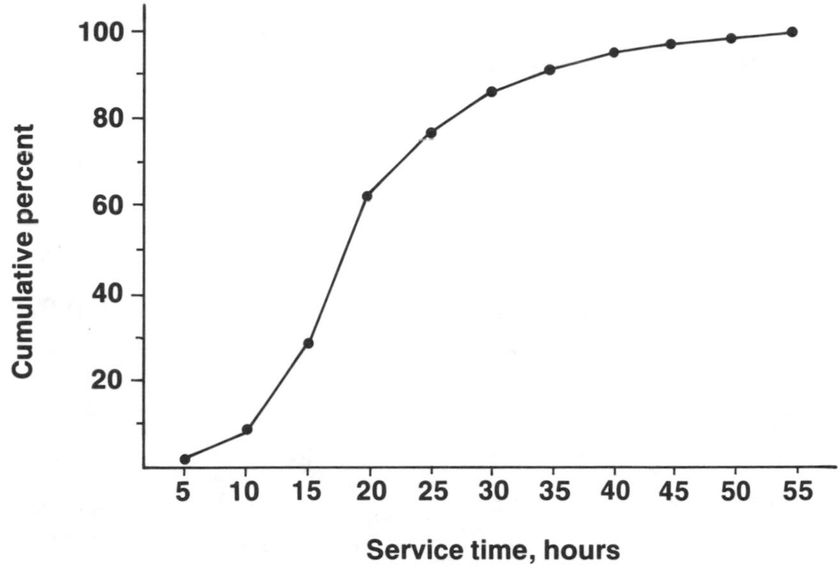

FIGURE 5.11

tribution sampling to another, weekly yields at a given manufacturing operation, or the quantity (dollar value) of monthly billings. For these types of data, since time is the gating factor, the trend chart would display the time frame and/or sequence of events along the horizontal axis and the data measure along the vertical axis.

To illustrate the construction and plotting of the trend chart, consider the error-free arrival quality data shown in Figure 5.12.

To construct the chart, the month interval is scaled along the horizontal axis with the percent error-free scaled along the vertical axis. Next, a dot corresponding to applicable percentages is plotted above each month (100% for January, 100% for February, 85.7% for March, etc.). The dots are then connected by drawing a straight line from one to the other. In actual practice, the dots would be plotted on a month-by-month basis as the data are received and computed.

The principal advantage of graphically displaying sequential data over time is the readily visible trend information it provides. For example, Figure 5.13 reflects an obvious negative trend in arrival quality which, on closer inspection of the data, could

Month	Quantity received	Quantity error free	Percent error free
January	5	5	100.0
February	2	2	100.0
March	7	6	85.7
April	4	4	100.0
May	10	9	90.0
June	14	12	85.7
July	22	19	86.3
August	28	22	78.5
September	19	17	89.4
October	31	25	80.6
November	33	22	66.6
December	45	28	62.2

FIGURE 5.12

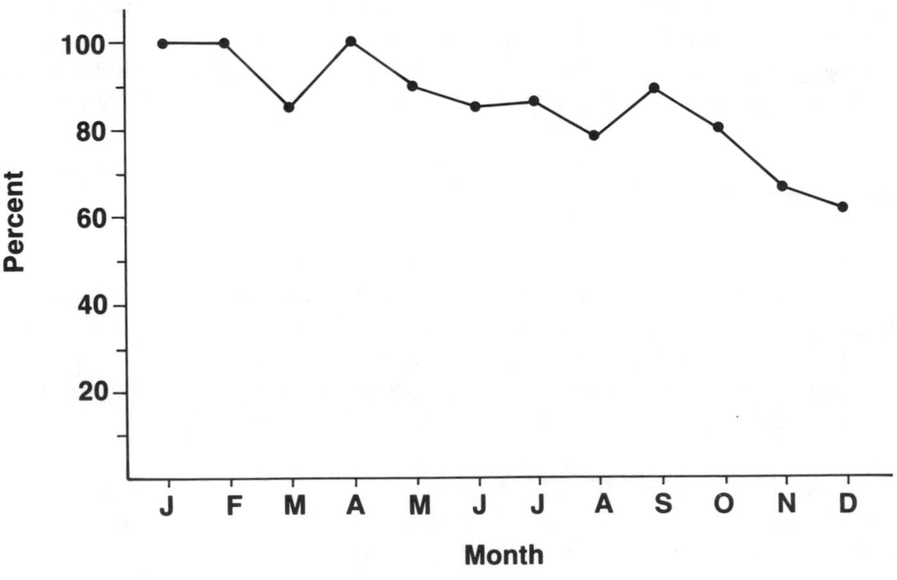

FIGURE 5.13

correlate with some correctable discrepancies. If so, then this trend information could alert management to the need for a fuller investigation of the data and subsequent corrective action measures.

BAR CHARTS

While both histograms and bar charts display data in the form of bars, the data depicted by the bars are quite different. For example, the bars of the bar chart are typically used to display the absolute value and/or relative relationships across several classification categories. This is quite different from the histogram where bars are used to depict cell interval and frequency within a single classification category.

To illustrate this difference, consider the previous procurement cycle histogram. Here, the single classification category is days, with the height and width of the bars representing the frequency of occurrence and cell interval. In this respect, the histogram displays some of the important and useful characteristics of the procurement cycle distribution (i.e., the range, central tendency, normality, or skewness of the data). While this information is important in understanding and displaying the distributional aspects of the procurement cycle data, or any other set of data, it fails to provide detailed information relative to why the distribution may be shaped in the manner it is (i.e., is the dispersion, skewness, or peakedness of the distribution a result of combining specific commodities, or a result of combining the data from the same commodity suppliers?). To answer these types of questions, the data should be arranged in a manner which highlights individual classification categories. For example, the bar chart shown in Figure 5.14 displays the various commodities associated with the procurement data. Presenting the data in this manner provides a clear picture of each commodity and its relative contribution to the overall procurement cycle.

Another common use of the bar chart is to display a single classification category over time. This is essentially the same approach as a trend chart except that instead of plotting and drawing lines between points, the height of the bars is used to depict applicable values.

When either bar or trend charts are used to display percentage data, caution should be exercised to assure that the quantities are sufficiently large enough to accurately compute and/or depict meaningful percentage values. For example, based on the bar chart in Figure 5.15, one could conclude that arrival quality significantly improved in March and significantly deteriorated in June.

One would probably not reach this conclusion, however, if he or she was aware that the percentage for March was based on the acceptance of a single product and the

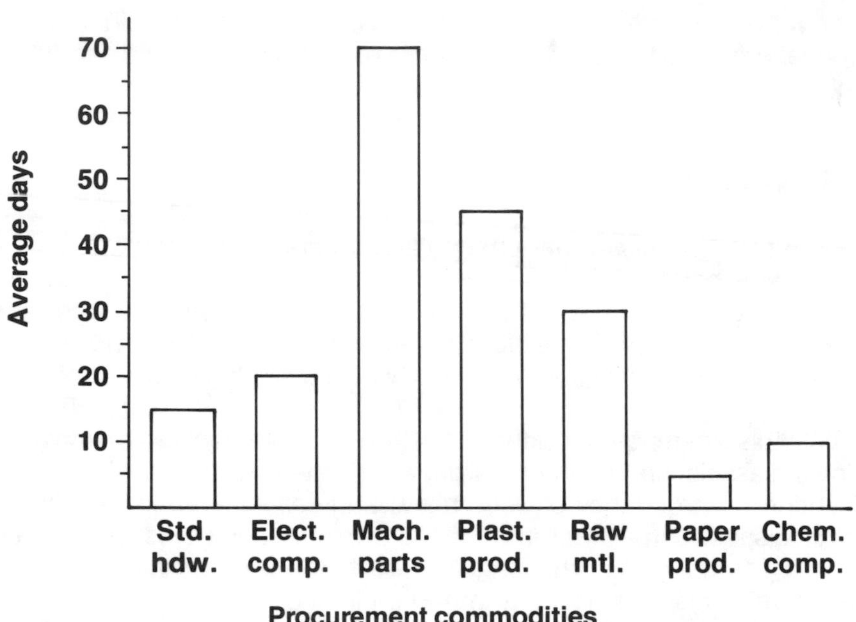

FIGURE 5.14

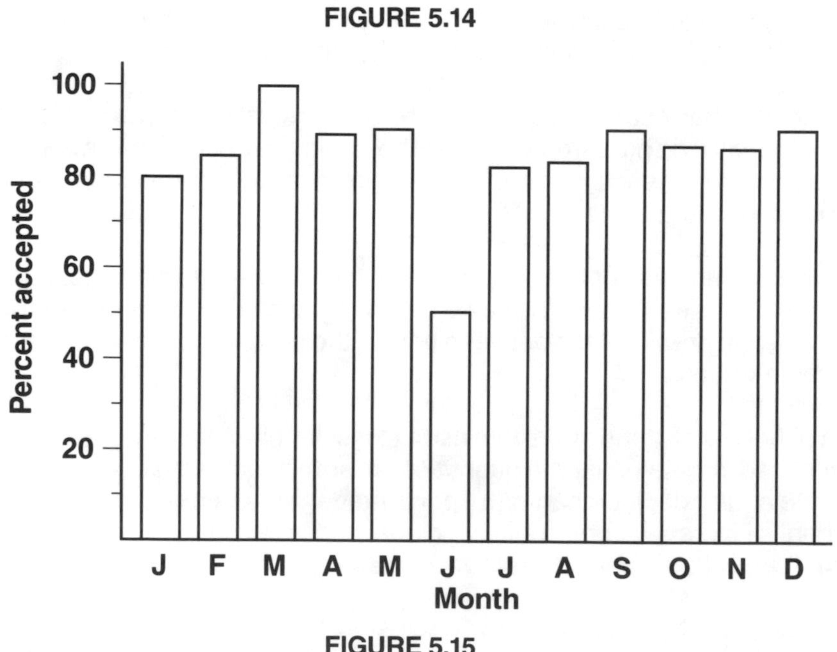

FIGURE 5.15

percentage for June based on the acceptance of one out of two products. To avoid these situations, bar charts should not be used to present data unless quantities are sufficiently large enough to accurately reflect meaningful differences among several classification categories or, as may often occur, the differences within a single category over time.

SCATTER CHARTS

While histograms, frequency polygons, and trend charts graphically summarize data along one dimension, for one variable, the scatter chart provides a graphic description of the relationship between two variables. That is, the data displayed in a scatter chart consist of two observations or measurements for each item or member within a group of items or members. For example, the measures of height and weight for each member in a group of people, years of schooling and income level for a group of adult males, or motivation versus productivity for a group of factory workers. Whatever the measures may be, the essential feature of the scatter chart is that each observation can be paired with another observation for each member of a group.

To illustrate this display of paired observations, consider the data representing the number of discrepant billing documents versus the number of documents processed (Figure 5.16).

Now, by plotting the pairs of scores on a graph, we have a visual representation of the relationship between these two variables (Figure 5.17).

In this case, there appears to be a fairly close positive relationship between the number of documents processed and the number of discrepant documents, which is determined by viewing both the slope and grouping of the plotted points. For example, Figure 5.18(a) is a graphical representation of a high positive relationship. In this case, the points fall very close to a straight line. This means that there is a strong relationship between the two variables. If the points were to fall exactly on a straight line, a perfect relationship would exist.

Note also that since high values of the X variable are associated with high values of the Y variable, a positive relationship exists. If high values of the X variable were associated with low values of the Y variable, as shown in Figure 5.18(b), then a negative relationship would exist between the two variables.

Figure 5.18(c) shows a relationship which is more or less random. In this case, no systematic tendency is observed for high values of the X variable to be associated with high values of the Y variable, or for low values of X to be associated with low values of Y.

Qty. proc.	Qty. discr.	Qty. proc.	Qty. discr.	Qty. proc.	Qty. discr.
300	4	100	1	200	1
450	4	200	3	550	7
300	2	400	3	350	4
400	5	150	2	250	3
550	8	550	5	450	6
500	6	600	8	600	7
300	3	350	3	400	4

FIGURE 5.16

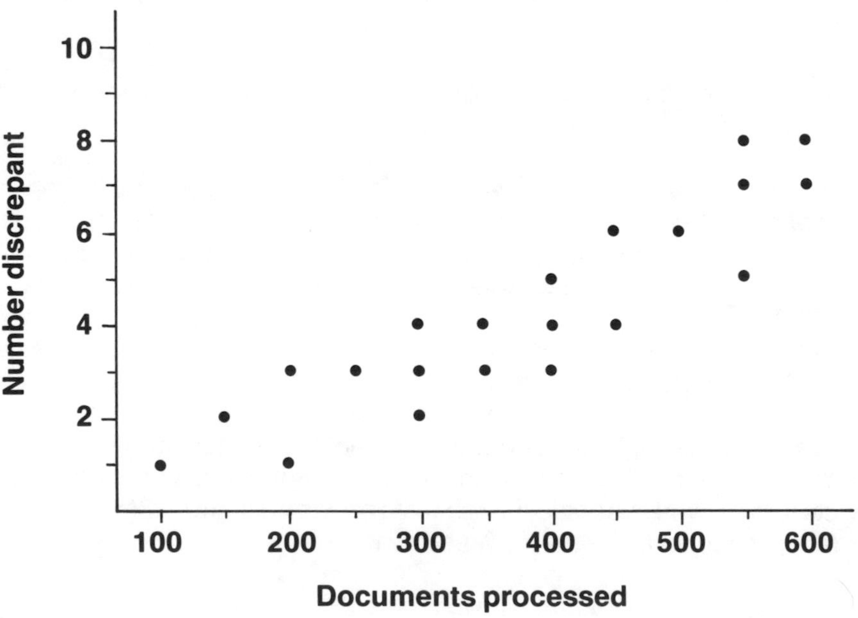

FIGURE 5.17

Between a perfect positive and perfect negative relation, an indefinitely large number of possible arrangement of points may occur. Accordingly, this represents an indefinitely large number of possible relations between two variables which, for clarification, can be quantified by computing a correlation coefficient.

The previous discussion has dealt with the linear relationship between two variables. There are other relationships, however, that are not linear. For example, Figure 5.18(d) represents a curvilinear relationship where the variables move in one direction and then shift to follow another direction. In this case, the influence of one variable may be positive up to a certain point and then become negative. One example of a curvilinear relationship would be between stress level and productivity. While increased stress may increase productivity up to a certain point, additional stress may tend to inhibit productivity.

When constructing a scatter chart, measures associated with frequency, percentage, errors, or any other measure of the dependent variable should be scaled along the vertical axis. The independent variables, such as total quantity, general performance, or time, are scaled along the horizontal axis. There are exceptions to this rule, but generally, the dependent variable is plotted on the Y axis and the independent variable is plotted on the X axis.

When we refer to two variables as independent and dependent, we are usually making an assumption that a cause-effect relationship exists between the two. This may not always be the case, however, and caution should be exercised when a scatter plot is used to predict one variable based on a knowledge of another. That is, scatter charts should not be used unless there is a very logical reason to suggest that a change in one variable will directly effect a change in the other, or unless one variable is deliberately manipulated to verify a corresponding change in the other. In all other circumstances, a scatter chart should only be used to simply evaluate or display the relationship between two related variables.

When a cause-effect relationship does exist, however, the variable causing the other to change is termed the independent variable. This is illustrated by the billing document example where the number of documents processed is the independent variable, and the number of discrepant documents the dependent variable. In this respect, if a true cause-effect relationship existed between the number of documents processed and the number of discrepant documents, it would be impossible to reduce the number of discrepant documents without reducing the number of documents processed (or developing more effective document processing procedures).

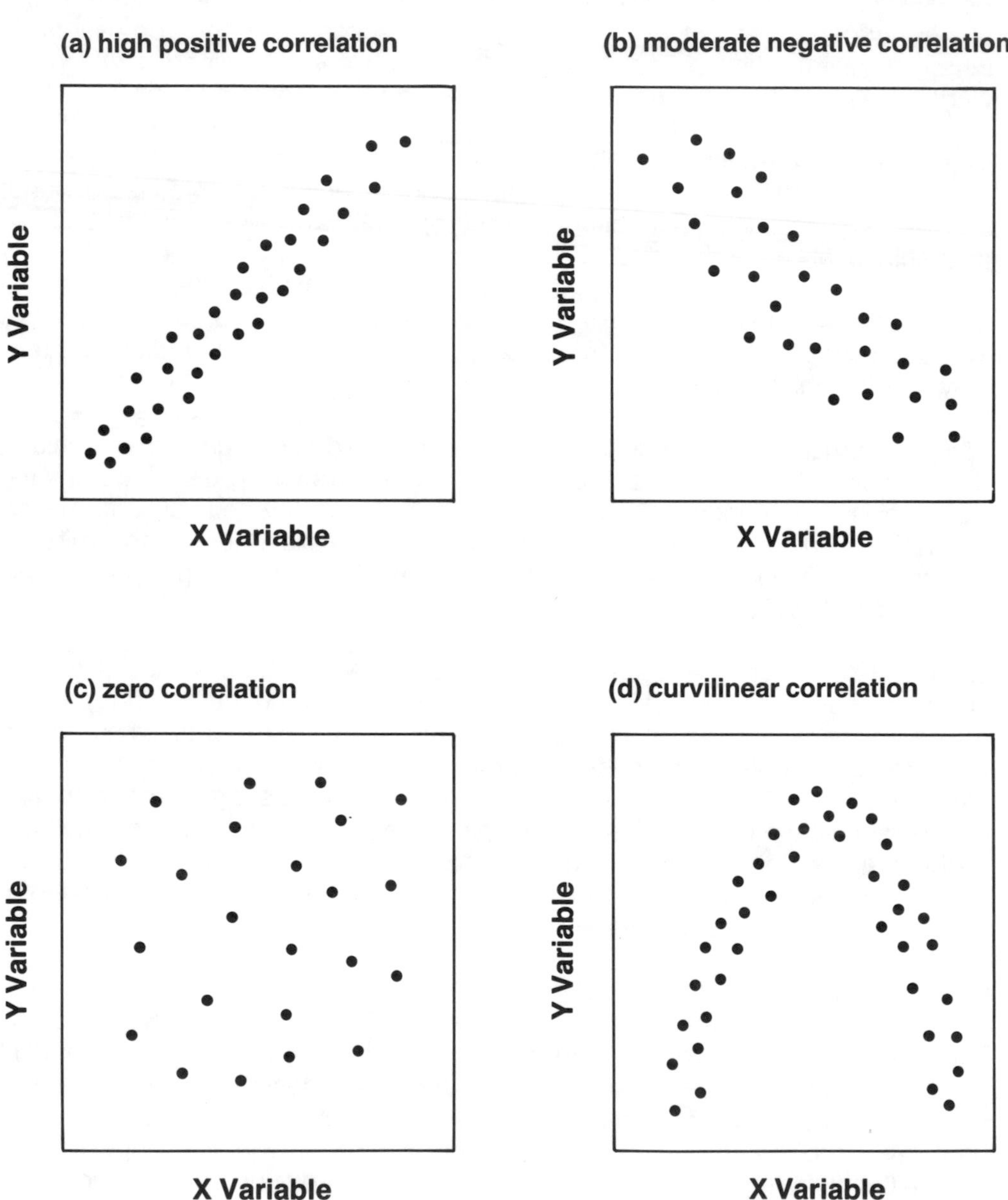
(a) high positive correlation
Y Variable
X Variable
(b) moderate negative correlation
Y Variable
X Variable
(c) zero correlation
Y Variable
X Variable
(d) curvilinear correlation
Y Variable
X Variable

FIGURE 5.18

SEQUENTIAL FREQUENCY CHART

For small sample quantities, or where quantities vary significantly from one report period to another, the sequential frequency chart is probably the most effective method for displaying trend information or conformance-nonconformance data. It is also quite effective in depicting subtle changes in both failure and rejection rates.

The sequential frequency chart is constructed by scaling total quantities along the horizontal axis and defect quantity along the vertical axis. Boundary limits also can be included to depict significant performance variations. When constructed in this manner, the chart is plotted by placing a dot at vector points corresponding to the total cumulative quantities processed and the total cumulative quantity defective. The dots are then connected with a solid line.

When scaling the chart, since both axes reflect total cumulative quantities, due consideration must be given to both the total expected quantities and the ability of the chart to reflect subtle changes within those quantity restraints.

To illustrate these considerations, as well as the construction and plotting of the chart, consider the following data used in graphing the previous arrival quality bar chart. For this illustration, monthly number defective and cumulative totals have been added to facilitate chart plotting (Figure 5.19).

The sequential frequency chart, in addition to desensitizing the effect of small sample percentages, provides a means for visually displaying acceptance sampling probabilities. For example, the goal/action lines in Figure 5.20 represent an acceptable quality level of 10%, a reject quality level of 30%, and alpha (α) and beta (β) risks of 5%. Thus, for cumulative points plotted in the goal area, there is a 95% probability that the product quality is equal to or better than 90% acceptable. Accordingly, there is a 95% chance that product quality levels are less than 70% acceptable when points are in the action area.

To construct an action-goal/accept-reject sequential frequency chart, the percent defective which is considered acceptable (AQL), the percent defective considered rejectable (RQL), and alpha (α) and beta (β) risks must be specified. The intersect points (h1, h2, f1, and f2) are then calculated using the equations shown at the end of this section.

The sequential frequency chart (Figure 5.21) can be used for many purposes. As a sequential f chart for reliability testing, as an action-goal trend chart for product inspections, as an indicator of significant changes in product/process quality, as a predictor of expected quality levels, or for most any other purpose where items, samples, time, or other characteristic classifications are accumulated in a sequential cumulative manner.

Month	J	F	M	A	M	J	J	A	S	O	N	D
Shipped	10	6	1	8	9	2	12	7	13	8	15	10
Accepted	8	5	1	7	8	1	10	6	12	7	13	9
Percent	80	83	100	87	88	50	83	85	92	87	86	90
Qty. def.	2	1	0	1	1	1	2	1	1	1	2	1
Cum. ship.	10	16	17	25	34	36	48	55	68	76	91	101
Cum. def.	2	3	3	4	5	6	8	9	10	11	13	14

FIGURE 5.19

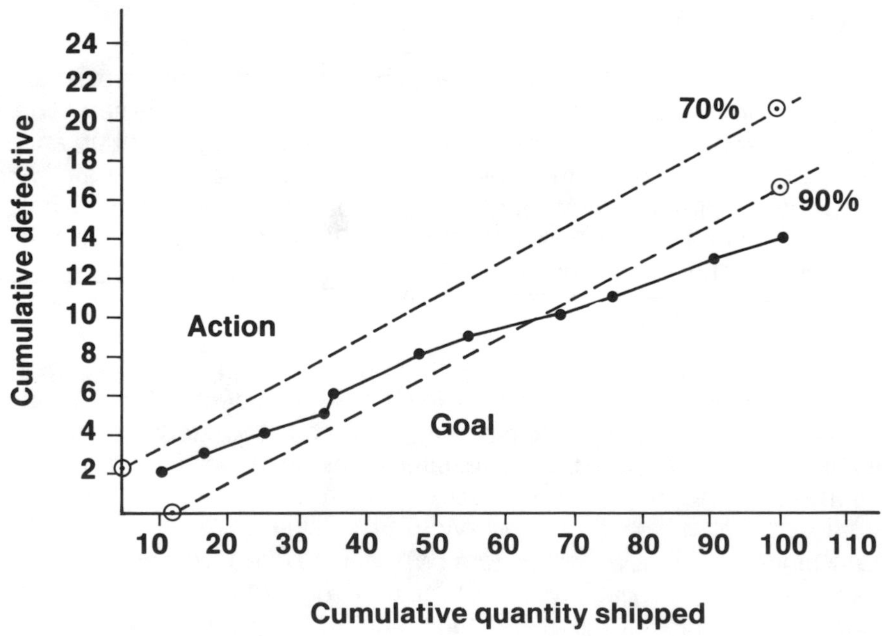

FIGURE 5.20

GRAPHICAL AIDS IN PROBLEM SOLVING

Effective problem solving is a four-step process of identifying the major problem, classifying its major contributors, verifying a cause-effect relationship, and eliminating or controlling the root-source. Accordingly, there are four graphical aids which facilitate this process: the Pareto chart, cause-effect diagram, scatter chart, and statistical control chart.

Pareto Chart

Pareto analysis consists of initially documenting the types and frequency of discrepancies encountered at a given operation or checkpoint, calculating the percentage in which each type of discrepancy contributes to the total, and then displaying that information in the form of a descending order histogram.

The first step in constructing a Pareto chart is to keep a tally score of the types of discrepancies encountered over a reasonable period of time, usually one week to one month. All discrepancies are then added together to arrive at a grand total. Finally, the percentage for each type of discrepancy is calculated and a descending order histogram constructed.

To illustrate this process, consider the tally mark distribution of product test discrepancies in Figure 5.22.

For these data, there are a total of 60 defects, with various quantities in each category. The percentage which each category contributes to the total, then, is calculated by dividing the number of defects in each category by the grand total. Note that an estimated repair cost for each category has been included as part of the tally distribution data. This additional information will be discussed later.

The Pareto chart is then constructed by listing the defect categories in descending order of magnitude along the X axis of the chart (Figure 5.23).

While this example relates to product testing, the same technique can apply to other operations: error rate among members of a typing pool, reasons for late reports from accounting, types of calls for field service, or reject rates by commodity code for purchase materials. Pareto analysis is useful in almost any endeavor where it is important to gain a focus on the major problem contributors (i.e., to separate the important few from the trivial many).

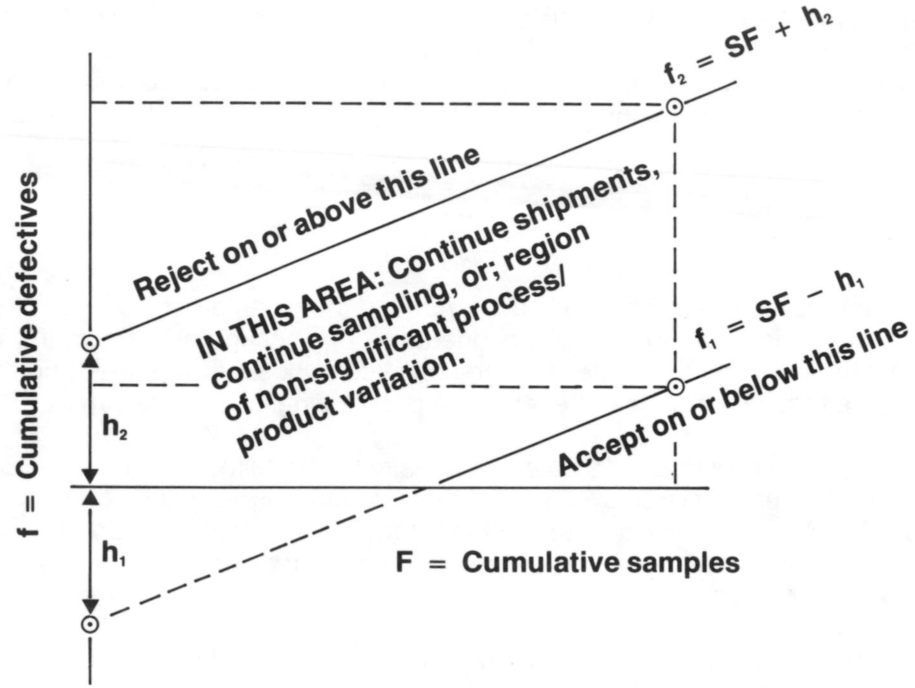

FIGURE 5.21

Defect category	Tally marks	Freq.	% of total	Repair cost
Volt. reg.	𝍸 𝍸 𝍸 𝍸 I	21	35.0	$1,000
Pulse amp.	𝍸	5	8.3	500
Rise time	𝍸 𝍸 IIII	14	23.3	2,500
Gain	IIII	4	6.6	4,000
Pulse width	𝍸 𝍸	10	16.6	1,500
Modulation	𝍸 I	6	10.0	750
Totals		60		$10,250

FIGURE 5.22

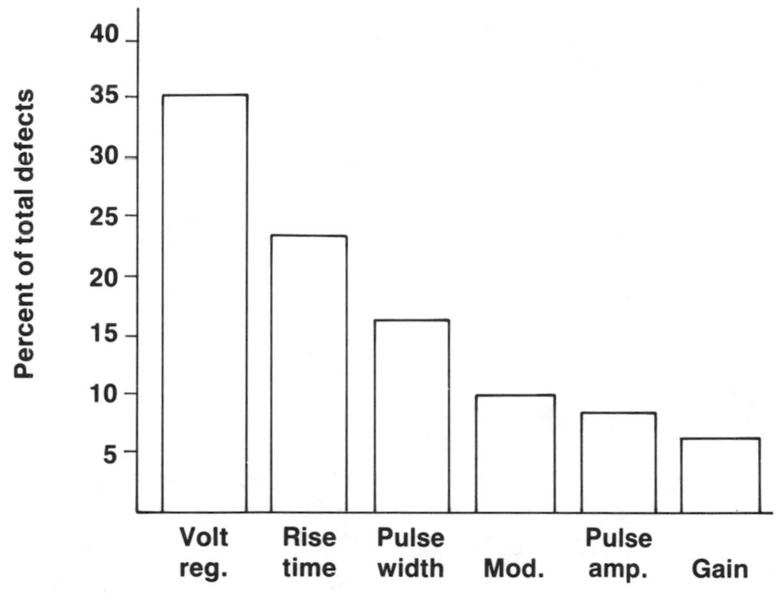

FIGURE 5.23

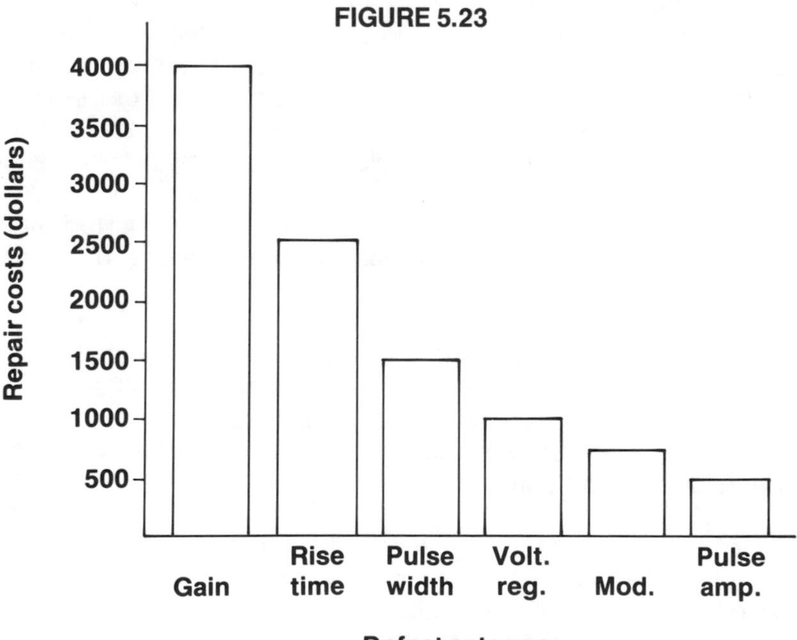

FIGURE 5.24

Returning to the listing of estimated repair cost on the tally mark distribution chart, it is important to remember that a single Pareto analysis may not always show the problem characteristic that needs to be corrected first. For example, voltage regulation is the most frequent reason for rejection in the tally mark distribution of defect categories. However, rework costs associated with voltage regulation are comparatively low, while the cost of reworking gain-related rejects is very high. From a cost standpoint, then, the first thing to do is to reduce the number of gain defects. Figure 5.24 is a Pareto chart of the same data as shown before. This time, however, the chart is constructed to show which corrective measure will have the greatest effect in reducing overall operational costs.

Cause-Effect Diagrams

Cause-effect diagramming is directly related to Pareto analysis. It is a structured technique for uncovering, layer-by-layer, the most probable causes (or sources of variation) associated with a defect category. Once the most probable causes are known, investigation can focus on determining the actual root causes.

To illustrate this procedure, consider the Pareto chart associated with the final testing of computer media disc shown in Figure 5.25. Using these data, the first step in constructing a cause-effect diagram is to determine the most probable causes for low amplitude (i.e., the major defect category). Opinions of technical staff members, preferably polled as a group, can greatly improve this process. For this illustration, however, we will simply interject coating thickness, orientation, and chemical mix properties as representing the most probable causes for low amplitude failures.

The structure of the cause-effect diagram, then, begins with low amplitude (the effect) which then branches to its three most probable causes: coating thickness, orientation, and mix properties (Figure 5.26).

Now follow the coating thickness branch. It expands with two probable causes of its own: coating operations (coaters) and polishing operations (polishers). Stemming from these are the next layers at probable causes. This process of identifying causes, layer-by-layer, continues until there is a comprehensive listing of all probable causes associated with low-amplitude failures.

Another form of the cause-effect diagram is called the "fishbone" diagram. While quite suitable for simple cause-effect analysis, fishbone diagramming may become confusing when several interacting relationships are involved. In those instances, the standard cause-effect layout, while perhaps not as graphic, may prove easier to construct and interpret (Figure 5.27).

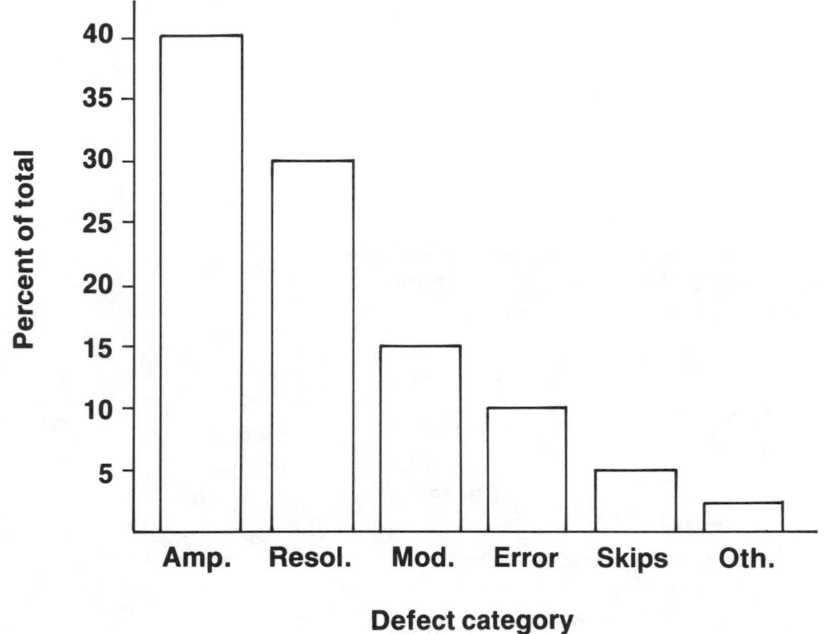

FIGURE 5.25

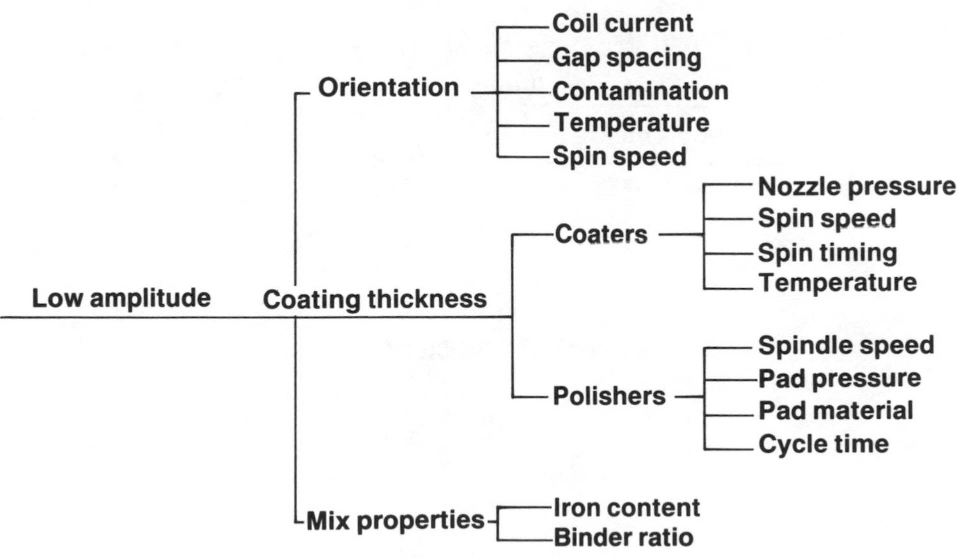

FIGURE 5.26

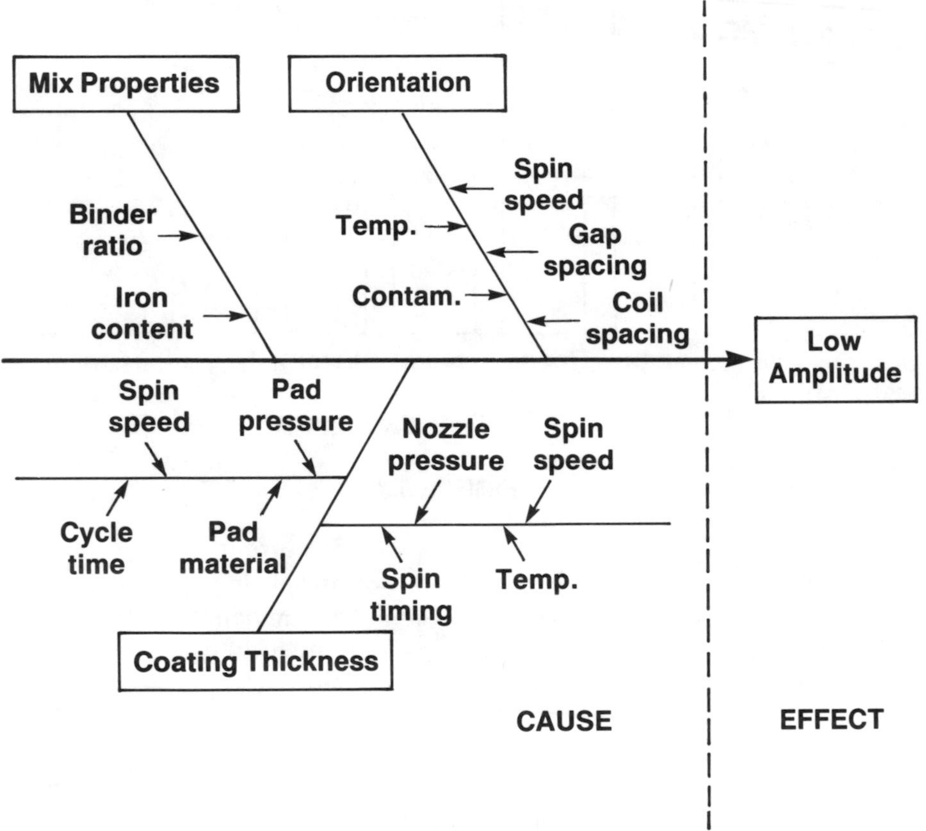
Mix Properties
Orientation
Binder ratio
Iron content
Spin speed
Temp.
Gap spacing
Contam.
Coil spacing
Low Amplitude
Spin speed
Pad pressure
Nozzle pressure
Spin speed
Cycle time
Pad material
Spin timing
Temp.
Coating Thickness
CAUSE
EFFECT

FIGURE 5.27

The cause-effect diagram is a road map. Its purpose is to chart routes for investigation when problems arise. The actual act of investigating problems, however, is a process of logical actions which lead the investigator to the root-cause source. In this sense, while the cause-effect diagram cannot solve the problem in and of itself, it can be quite useful in helping to reduce the number of possible cause categories to be investigated.

Scatter Charts

A scatter chart provides a quick assessment of a probable relationship between two variables. As such, it can provide a quick assessment of the probable relationship between an identified cause characteristic and a problem effect. For example, coating thickness was identified as a probable cause resulting in low-amplitude discs. A scatter chart would prove this relationship true or false. If it were true, as reflected by a tight grouping of data points, then further investigations could be directed toward the major characteristic, or characteristics, affecting coating thickness. If the chart did not reflect this relationship, however, then investigations could focus on more productive probable causes; thereby saving a great deal of time and effort that could have been spent in investigating the various process associated with this characteristic.

For this very reason, it is important to not only verify that a strong relationship exists between the two variables, but that it is a cause-effect relationship.

For example, if a cause-effect relationship exists between coating thickness and amplitude, then an increase in coating thickness (the independent variable) should result in a corresponding increase in amplitude (the dependent variable). Accordingly, a decrease in amplitude should result in a corresponding decrease in coating thickness, assuming that the cause-effect relationship is positive. Plotting these two conditions on a scatter chart would reflect whether or not a true cause-effect relationship existed between these variables.

ROOT-CAUSE ANALYSIS

The fourth and last step in effective problem solving is to either eliminate or control the problem at its root-source. This is actually a two-step process. First, the root-source must be identified. Thereafter, once the root-cause is uncovered and a cause-effect relationship verified, actions can begin to either eliminate the cause at its source, or to control its effect. In the first case, it is often possible to eliminate the cause through design changes, or through changes in manufacturing and/or screening procedures. When this approach is not practical, the next best solution may be the application of statistical control charts. This approach, as explained in Section 4, is a process of systematically eliminating the symptom effect by continuously reducing the inherent variability of the cause characteristic.

So far, we have illustrated individual component problems where it is possible to go from a single Pareto chart directly into cause-effect diagramming. At the final product level, however, where thousands of components can be involved, it may be impossible to display all possible failure mechanisms in a single cause-effect diagram. Therefore, it is customary to prepare at least three Pareto charts: one which depicts the reason for rejection at the system or final test level, a second which depicts the subassembly and/or component(s) associated with the major system failure, and a third which depicts the failure categories associated with the major defective component. From that point, subsequent Pareto charting, cause-effect diagramming, and scatter plotting would continue until the root-cause is uncovered. In this manner, root-cause analysis is like peeling an onion one layer at a time.

The graphical aids and problem-solving techniques presented in this section are universal in that they provide a structure for identifying and assessing probable causes of problems found in engineering, accounting, marketing, procurement, or any other business discipline. For example, consider the case of a regional marketing manager whose goal is to increase sales by 10%. For this purpose, a Pareto chart depicting individual sales performance would aid in identifying the person(s) who could contribute the most toward increasing overall performance (i.e., trying to increase the sales quota of a high performer would probably yield less overall gain than increasing the sales volume of a low performer).

After preparing a Pareto chart to identify those individuals who are performing less well than others, the next step is to construct a cause-effect diagram to identify applicable performance variables. Then, by comparing poor performers on a variable-by-variable basis to the group average, specific areas for improvement become more apparent.

CONVENTIONS FOR CONSTRUCTION OF GRAPHS

For clarity and ease of interpretation, the following rules should be followed when preparing graphs, charts, or illustrations:

1. Graphing of frequency distributions should be such that the horizontal axis represents scores while the vertical axis represents frequencies.
2. Graph arrangement should be from left to right. The low numbers on the horizontal scale should be on the left, and the low numbers on the vertical scale should be toward the bottom.
3. For aesthetic reasons, scale units should be selected such that the ratio of graph height to length is approximately 3:5.
4. Whenever possible, the vertical axis should be scaled such that the 0 point falls at the point of intersection with the horizontal axis. When this is not practical due to data values, it is customary to designate the point of intersection as the 0 point, make a small break in the vertical axis, and then scale in accordance with the data.
5. Where applicable, both axes should be clearly labeled to reflect both the quantity measured and the measurement unit.
6. Numbering and lettering of scale points should be horizontal for both axes. Legends should be parallel to their particular axes.
7. While legibility should be used as a guide in determining the number of curves to place on a graph, a good general rule is to use no more than four.
8. Use distinct geometric forms, such as circles or squares, for plot points.
9. When considering scale values, the range and/or separation between grid marks should be selected so that plotted curves span the entire graph.
10. Indicate scale values by grid marks placed on each axis at the appropriate intervals.
11. When applicable, enclose figures with a box outline to distinguish them from the text.
12. Every graph should be assigned a descriptive title which states precisely what it is about.

SEQUENTIAL FREQUENCY CHART CONSTRUCTION

To construct a sequential frequency chart, one must first establish the acceptable/rejectable quality levels (P1 and P2) as well as the associated risk factors (α and β). From there, the slope and intersect point for the accept/reject lines (f1 and f2) can be computed after first completing the necessary preliminary calculations. To demonstrate this series of events, the following represents the assumptions and calculations used in constructing the previously illustrated Arrival Quality Chart (Figure 5.19, page 100).

P_1 = .10 The percentage of defective products which, on the average, is considered to be an acceptable level of quality or, for the Arrival Quality Chart, a reasonable goal to be achieved.

P_2 = .30 The level of quality considered unacceptable and/or the threshold for corrective action measures.

α = .05 The "alpha" risk associated with an incorrect judgment at the P_1 level of quality.

β = .05 The "beta" risk associated with an incorrect judgment at the P_2 level of quality.

$$g_1 = \log\frac{P_2}{P_1} = \log\frac{.3}{.1} = \log 3 = .477$$

$$g_2 = \log\frac{1 - P_1}{1 - P_2} = \log\frac{.9}{.7} = \log 1.285 = .109$$

$$a = \log\frac{1 - \beta}{\alpha} = \log\frac{.95}{.05} = \log 19 = 1.278$$

$$b = \log\frac{1 - \alpha}{\beta} = \log\frac{.95}{.05} = \log 19 = 1.278$$

$$h_1 = \frac{b}{g_1 + g_2} = \frac{1.278}{.477 + .109} = 2.180$$

$$h_2 = \frac{a}{g_1 + g_2} = \frac{1.278}{.477 + .109} = 2.180$$

$$s = \frac{g_2}{g_1 + g_2} = \frac{.109}{.477 + .109} = .186$$

$$f_1 = s F - h_1$$

For F = 0: (.186) (0) − 2.180 = −2.180
For F = 100: (.186) (100) − 2.180 = 16.420

$$f_2 = s F + h_2$$

For F = 0: (.186) (0) + 2.180 = 2.180
For F = 100: (.186) (100) + 2.180 = 20.780

Note: when $\alpha = \beta$, then $h_1 = h_2$

An alternate method of calculating f_1 is to solve for the cumulative quantity where f_1 intersects the X axis. In this case:

$$F = \frac{h_1}{s} = \frac{2.180}{.186} = 11.7$$

Based on the previous calculations, h_1, or the goal line, intersects the X axis at a cumulative quantity of approximately 12 units. This is the minimum number of continuously acceptable samples necessary to base a judgment that the goal has been achieved. On the other hand, 3 consecutive defective units would constitute a 95% probability that the overall arrival quality is equal to or less than 70% acceptable (i.e., 30% defective). Accordingly, if the number of defectives is between the action and goal lines, a 95% probability statement cannot be made as to the predicted or expected arrival quality level. In this case, additional units and/or defectives are necessary before a firm judgment can be made.

Exercise Worksheets

Exercise 5.1

PROBLEMS

1. Using a class interval of 5 for the following data, prepare a frequency distribution listing class interval values, frequency of occurrence, exact limits, and interval midpoints.

23	29	18	17	36
29	29	22	33	26
39	17	38	44	23
22	9	40	14	33
21	23	28	1	10
21	7	29	39	33
31	11	30	12	26
49	28	23	43	5
24	19	32	25	16
15	25	34	24	26

Class interval	Frequency	Exact limits	Midpoints

Exercise 5.1 (continued)

2. Prepare a histogram for the data in Problem 1.

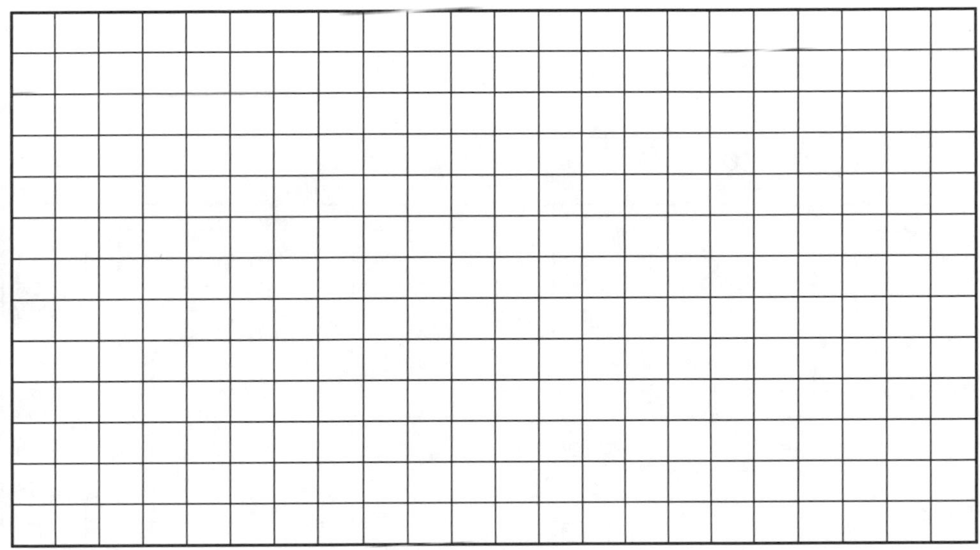

3. Prepare a frequency polygon for the data in Problem 1.

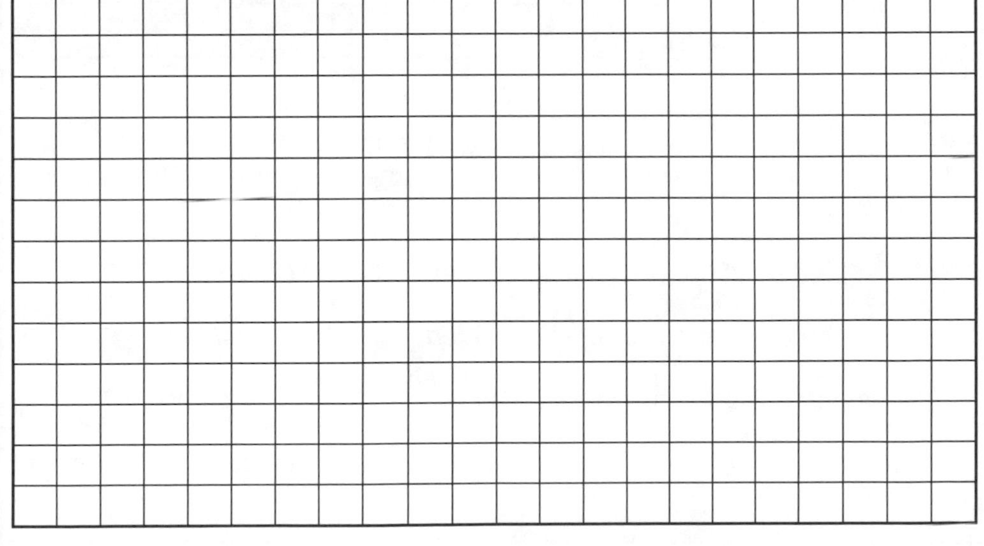

Exercise 5.1 (continued)

4. Prepare a cumulative percent polygon for the following data:

2	11	6	4	18	1	9	3	4	15
8	16	12	11	16	5	5	7	5	7
11	9	5	15	14	13	6	9	7	7
7	10	10	8	13	12	8	10	10	12
8	10	9	9	7	9	10	12	10	12

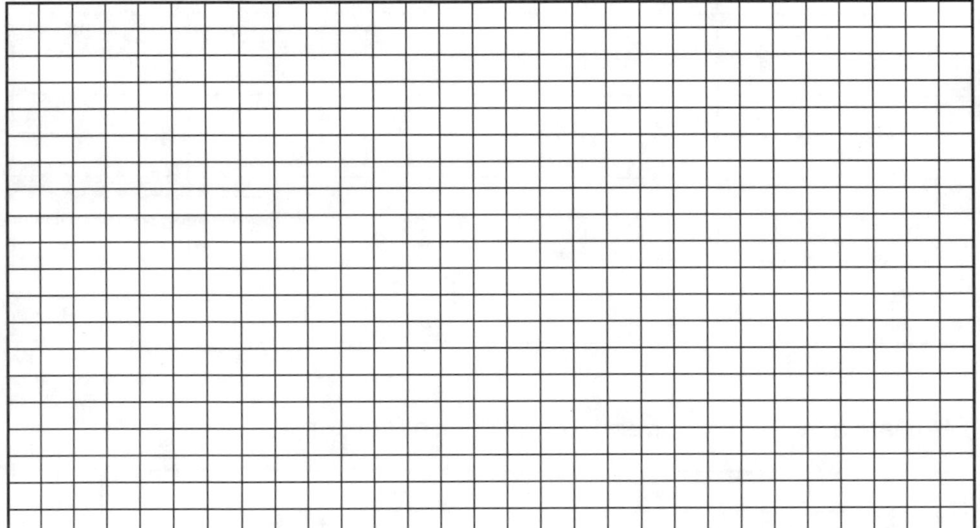

5. Assuming the cumulative percent polygon prepared in Problem 4 represents the time to complete a problem fix at a customer's site, answer the following questions:

   a) What percent of the fixes are completed within five hours?

   b) What percent of the fixes are completed within 10 hours?

   c) What percent of the fixes take greater than 12 hours to complete?

   d) How many hours does it take to complete approximately 90% of the site fixes?

SECTION 6

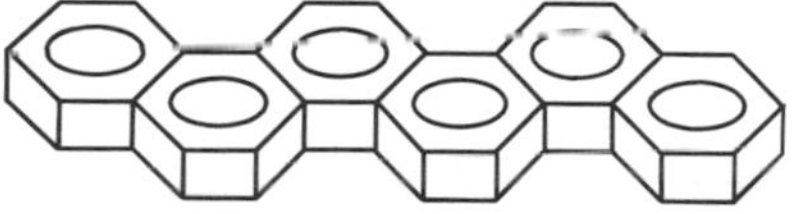

CORRELATION AND REGRESSION

CORRELATION

Correlation is concerned with the relationship between two sets of measurements (x and y).

Correlation studies are performed to first determine if a relationship exists between two variables and then, to determine if one variable can be estimated from knowledge of the other (i.e., if we know something about variable x, can we predict something about variable y?).

Correlation, or the relationship between two variables, can be graphically represented in the form of a scatter chart. Figure 6.1 is a graphical representation of a high positive relationship where high values of x are associated with high values of y. Figure 6.2 shows a low positive relationship. Figure 6.3 shows a random relationship where there is no particular tendency for high values of x to be associated with high values of y. Figure 6.4 shows a fairly high negative relationship where high values of x are associated with low values of y.

CORRELATION COEFFICIENT

The correlation coefficient, ranging in value from -1 to $+1$, provides a means of quantifying the relationship between two variables.

A value of -1, graphically illustrated when all points lie on a descending straight line, describes a perfect negative relationship where y proportionally decreases as x increases.

A value of $+1$, graphically illustrated when all points lie on an ascending straight line, describes a perfect positive relationship where y proportionally increases as x increases.

If the correlation coefficient is 0, no systematic tendency is observed between the value of x and y. The points may be spread uniformly over the xy plane, lie on a horizontal plane, or have a pattern that is symmetrical about a horizontal or vertical line (Figure 6.5).

For practical application in a manufacturing environment, a correlation coefficient of .80 or better is considered sufficient to consider that a strong relationship exists between two variables.

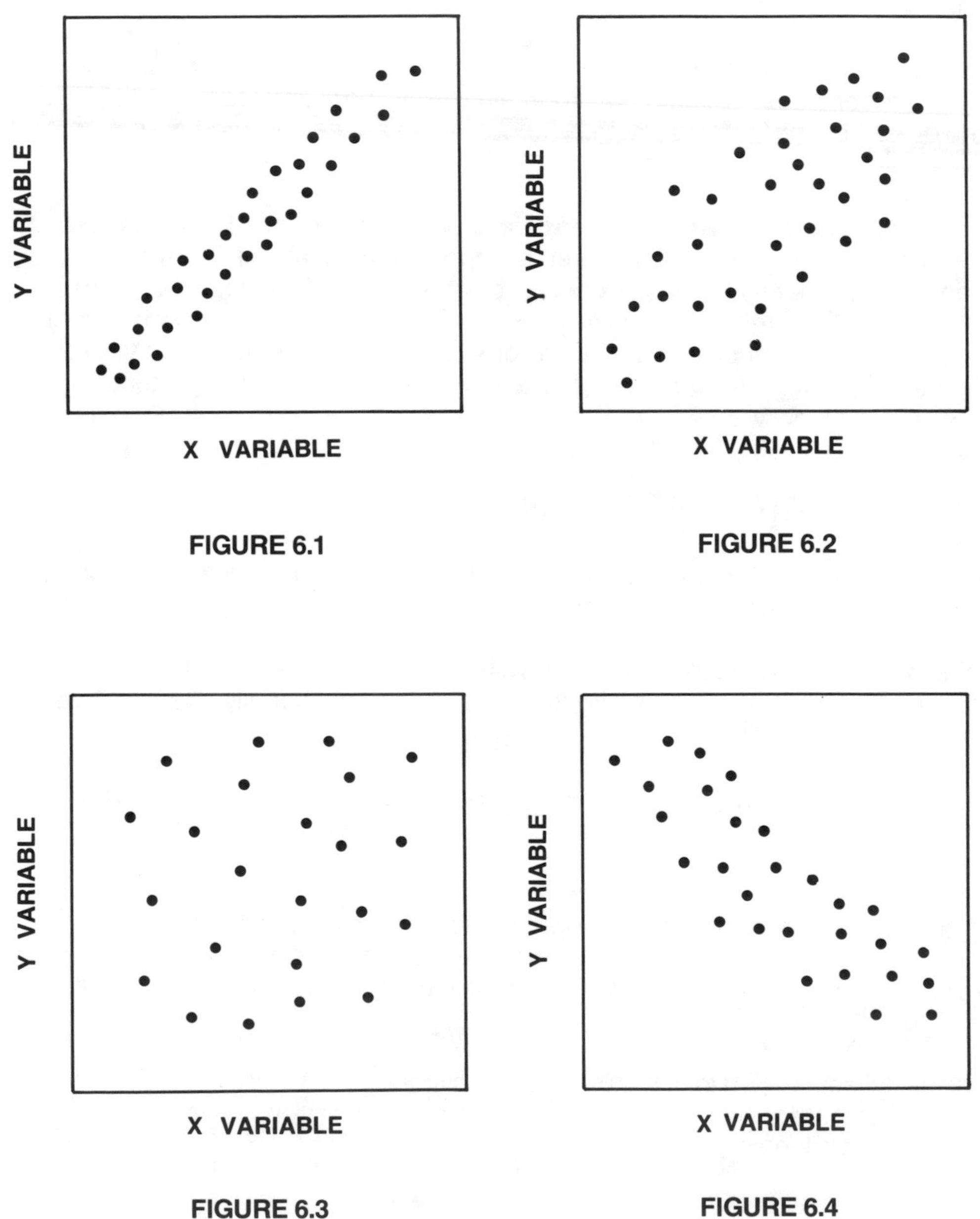

FIGURE 6.1

FIGURE 6.2

FIGURE 6.3

FIGURE 6.4

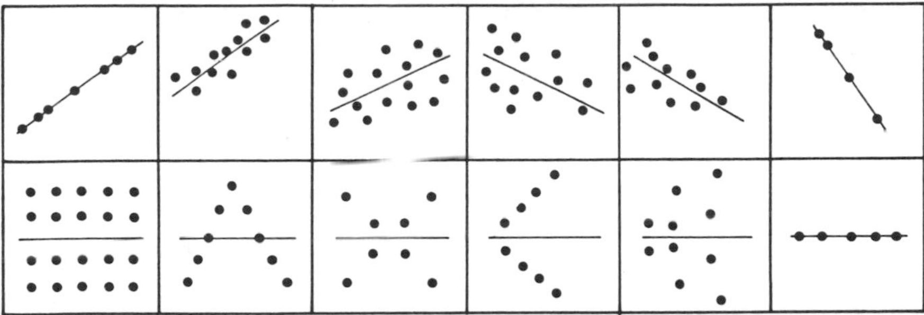

FIGURE 6.5

The correlation coefficient is obtained by converting actual measures to z score values, multiplying the paired scores, summing the results, and dividing that total by n − 1

$$r = \frac{\Sigma\, z_x z_y}{n - 1}$$

where r is the correlation coefficient, z_x is the z score for the x variable, z_y is the z score for the y variable, and n is the number of paired scores.

This method of calculating the correlation coefficient is somewhat laborious because it requires the conversion of all x and y values to z scores. However, since $z_x = (x - \bar{x})/s_x$ and $z_y = (y - \bar{y})/s_y$ a more convenient formula is:

$$r = \frac{\Sigma\, D_x D_y}{\sqrt{(\Sigma D_x^2)\,(\Sigma D_y^2)}}$$

where D_x is the deviation of the x values from $\bar{x}$ and D_y is the deviation of the y values from $\bar{y}$.

For ease of calculation, it also is convenient to arrange the respective numerical values in columns — much in the same fashion as calculating standard deviation. The first two columns contain the paired measures of x and y. These columns are summed and divided by n to obtain the values of $\bar{x}$ and $\bar{y}$. Column 3 (D_x) contains the deviations of x from $\bar{x}$ $(x - \bar{x})$, and column 4 (D_y) contains the deviations of y from $\bar{y}$ $(y - \bar{y})$. Columns 5 and 6 (D_x^2 and D_y^2) contain the squares of these deviations which are summed to obtain the $\Sigma\, D_x^2$ and $\Sigma\, D_y^2$ values.

Column 7 (D_xD_y) contains the products of D_x and D_y, which are summed to obtain the $\Sigma\ D_xD_y$ value (Figure 6.6).

	1	2	3	4	5	6	7
ITEM	x	y	D_x	D_y	D_x^2	D_y^2	$D_x D_y$
1	5	1	−1	−3	1	9	+3
2	10	6	+4	+2	16	4	+8
3	5	2	−1	−2	1	4	+2
4	11	8	+5	+4	25	16	+20
5	12	5	+6	+1	36	1	+6
6	4	1	−2	−3	4	9	+6
7	3	4	−3	0	9	0	0
8	2	6	−4	+2	16	4	−8
9	7	5	+1	+1	1	1	+1
10	1	2	−5	−2	25	4	+10
	60	40	0	0	134	52	48
	$\overline{X} = 6.0$	$\overline{y} = 4.0$			ΣD_x^2	ΣD_y^2	$\Sigma D_x D_y$

$$r = \frac{\Sigma D_x D_y}{\sqrt{(\Sigma D_x^2)(\Sigma D_y^2)}} = \frac{48}{\sqrt{(134)(52)}} = +.58$$

FIGURE 6.6

PREDICTION

Prediction and correlation are closely related topics and an understanding of one requires an understanding of the other. The presence of a 0 correlation between two variables x and y can usually be interpreted to mean that they bear no systematic relation to each other. A knowledge of x tells us nothing about y, and a knowledge of y tells us nothing about x. Thus, in predicting x from y or y from x, no prediction better than a random guess is possible.

The presence of correlation between x and y however, implies that if we know something about x we know something about y, and vice versa. Accordingly, if knowledge about x implies some knowledge about y, a prediction about y from x is possible. Additionally, the greater the absolute value of correlation between x and y, the more accurately we can predict the value of one variable by knowing the value of the other. If the correlation between the two variables is either -1 or $+1$, a perfect prediction is possible.

REGRESSION

A straight line between any two points can be represented by the equation: $y = a + bx$ where (a) is the intercept point where the line crosses the y axis, where (b) is the slope of the line (the number of units that y increases when x increases one unit), and where x and y are variables. Thus, only two points are needed to draw a straight line.

In predicting one variable based on a knowledge of another, the equation for a straight line which best "fits" all the data (since several data points are needed to make an accurate prediction) is called the regression line. The regression line of y on x is used to predict y based on a knowledge of x and the regression line of x on y is used to predict x based on a knowledge of y.

This method of fitting a regression line to a set of data containing two variables is called the "method of least squares" (Figure 6.7).

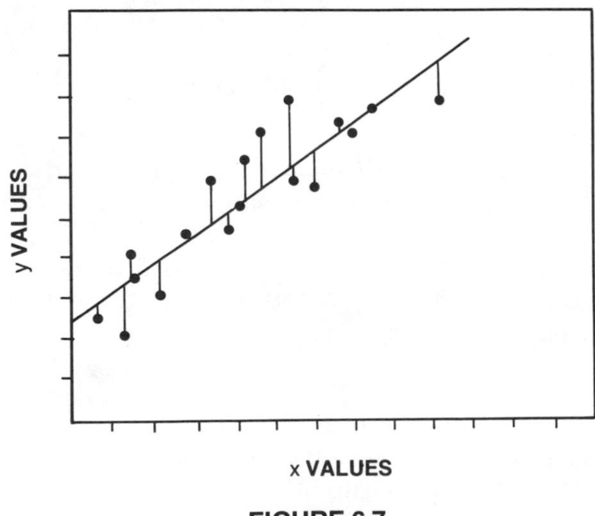

FIGURE 6.7

Regression Without the Correlation Coefficient

If the correlation coefficient is not computed prior to calculating the regression line, the formula for the regression line of x on y is:

$$x' = b_{xy}\, y + a_{xy}$$

where the symbol x' denotes the predicted value of x and where b_{xy} and a_{xy} (the slope of the line and the intercept point) are calculated using the following formulas:

$$b_{xy} = \frac{\Sigma\, xy - (\Sigma x \Sigma\, y/n)}{\Sigma\, y^2 - [(\Sigma\, y)^2/n]}$$

$$a_{xy} = \frac{\Sigma\, x - b_{xy}\, \Sigma\, y}{n}$$

Regression Using the Correlation Coefficient

Since there is a direct relationship between the correlation coefficient and the regression line, either x on y or y on x, the product of b_{xy} times b_{yx} results in the same value as obtained by squaring the correlation coefficient (r^2). For this reason, the following simplified formulas are usually used to compute regression lines:

$$y' = r \times \frac{s_y}{s_x}(x - \bar{x}) + \bar{y}$$

$$x' = r \times \frac{s_x}{s_y}(y - \bar{y}) + \bar{x}$$

where s_y and s_x are the standard deviation values for the y and x measures, where $\bar{y}$ and $\bar{x}$ are the averages for the y and x measures, and where x and y are selected points along the x and y axis. For example, since two points are required to draw the regression line, two values of x (usually the lowest and highest values in the x measures are selected when predicting y based on x; or the lowest and highest values of y when predicting x based on y).

Example 6.1
Given the information on the following page, compute a regression line for y on x (i.e., predicted value of y based on a knowledge of x).

Variable	Mean	S	Low Value	High Value	Corr. Coef.
x	112.44	9.04	98	131	.89
y	64.16	6.74			

For x = 98: $y' = r \times \frac{s_y}{s_x}(x - \bar{x}) + \bar{y}$

$= .89 \times \frac{6.74}{9.04}(98 - 112.44) + 64.16$

$= 54.5$

For x = 131: $y' = .89 \times \frac{6.74}{9.04}(131 - 112.44) + 64.16$

$= 76.4$

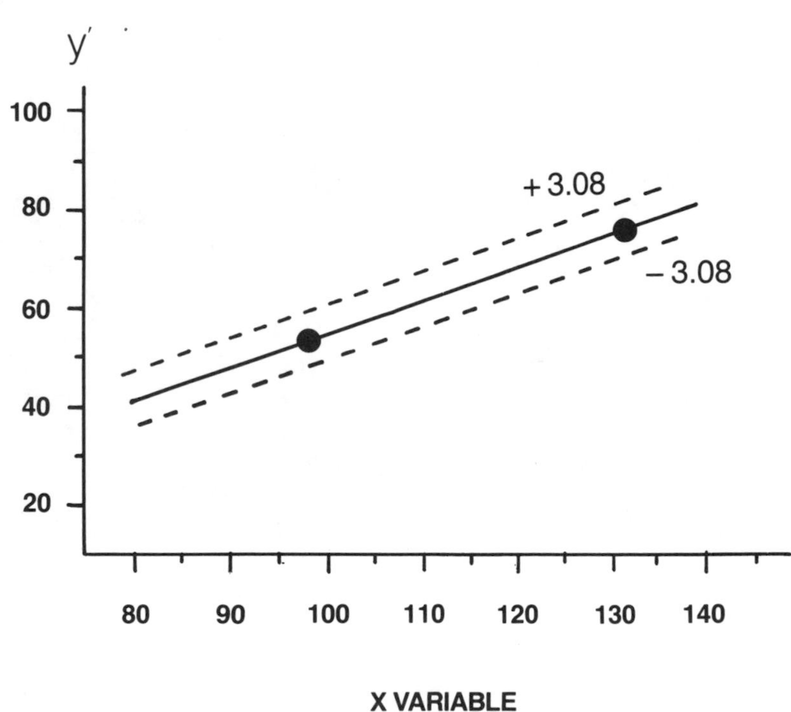

STANDARD ERROR OF ESTIMATE

Unless all paired observations fall directly along a straight line, a coefficient of +1 or −1, there is some inherent error in predicting one variable based on another. The error of the estimate is defined algebraically as $\Sigma\,(y - y')^2/n - 1$ for y on x and $\Sigma\,(x - x')^2/n - 1$ for x on y. When computing regression based on the correlation coefficient, the following simplified formulas can be used:

$$s_{yx} = s_y\sqrt{1 - r^2} \text{ and } s_{xy} = s_x\sqrt{1 - r^2}$$

For the previous example, the y′ estimate error is:

$$s_{yx} = 6.74\sqrt{1 - (.89)^2} = 3.08$$

Exercise Worksheets

Exercise 6.1

CORRELATION AND REGRESSION

PROBLEM: Several final test failures are attributable to a particular subassembly type. Investigation at the supplier's facility reveals that a different type of equipment is now being used for acceptance testing of the subassembly. Given the following 10-piece correlation sample measurements, listed on the correlation worksheet as x for your equipment and y for the supplier's equipment, what action would you recommend to assure receipt of acceptable products?

STEP 1: Calculate the required values on the correlation worksheet and then, using the following formula, calculate the correlation coefficient (r).

$$r = \frac{\Sigma D_x D_y}{\sqrt{(\Sigma D_x^2)(\Sigma D_y^2)}}$$

$$= \frac{[\quad\quad]}{\sqrt{[\quad\quad][\quad\quad]}} = [\quad\quad]$$

Exercise 6.1 (continued)

CORRELATION WORKSHEET

x VALUES	y VALUES	D_x $(x - \bar{x})$	D_y $(y - \bar{y})$	D_x^2	D_y^2	$D_x D_y$
118	66	[ ]	[ ]	[ ]	[ ]	[ ]
99	50	[ ]	[ ]	[ ]	[ ]	[ ]
118	73	[ ]	[ ]	[ ]	[ ]	[ ]
121	69	[ ]	[ ]	[ ]	[ ]	[ ]
123	72	[ ]	[ ]	[ ]	[ ]	[ ]
98	54	[ ]	[ ]	[ ]	[ ]	[ ]
131	74	[ ]	[ ]	[ ]	[ ]	[ ]
121	70	[ ]	[ ]	[ ]	[ ]	[ ]
108	65	[ ]	[ ]	[ ]	[ ]	[ ]
111	62	[ ]	[ ]	[ ]	[ ]	[ ]
$\bar{x}$ [ ]	$\bar{y}$ [ ]	S_x [ ]	S_y [ ]	ΣD_x^2 [ ]	ΣD_y^2 [ ]	$\Sigma D_x D_y$ [ ]

Exercise 6.1 (continued)

STEP 2: Enter specified values from correlation worksheet.

Lowest x value	$\bar{x}$	$\bar{y}$
[ ]	[ ]	[ ]
Highest x value	s_x	s_y
[ ]	[ ]	[ ]

STEP 3: Calculate low and high regression points for y:

$$y' = r\left[\frac{s_y}{s_x}(x-\bar{x})\right] + \bar{y}$$

Low Point:

$$y' = [\quad] \times \left[\frac{[\quad]}{[\quad]}\left([\quad] - [\quad]\right)\right] + [\quad]$$

$$= [\quad]$$

High Point:

$$y' = [\quad] \times \left[\frac{[\quad]}{[\quad]}\left([\quad] - [\quad]\right)\right] + [\quad]$$

$$= [\quad]$$

Exercise 6.1 (continued)

STEP 4: Calculate the standard error of y on x.

$$S_{yx} = S_y \sqrt{1 - r^2}$$

$$= [\quad] \sqrt{1 - [\quad]^2} = [\quad]$$

STEP 5: Plot regression line with upper and lower error limits.

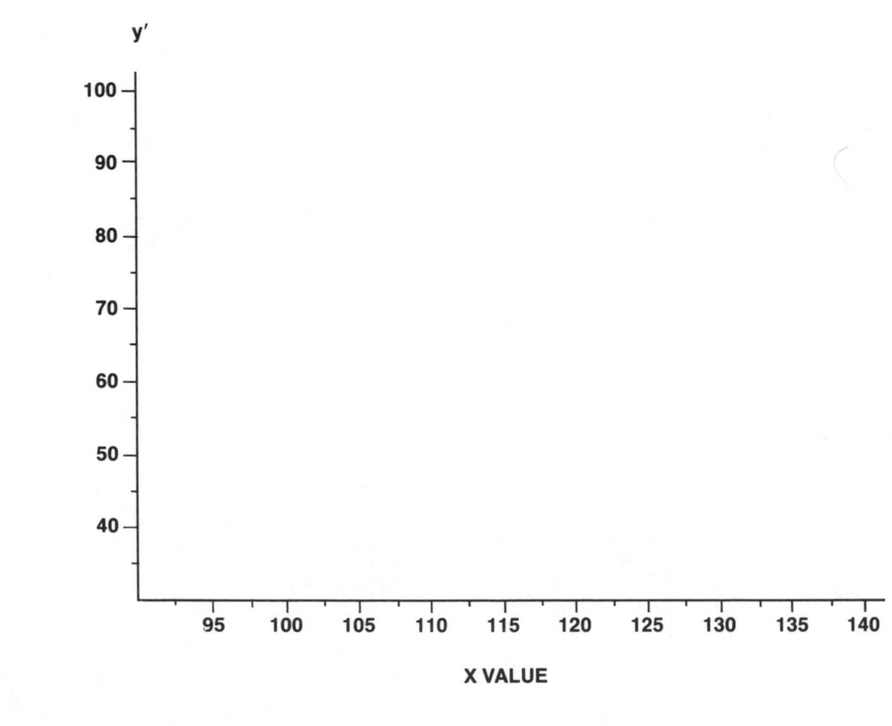

SECTION 7

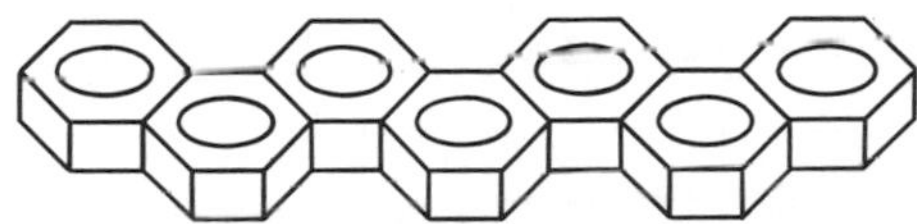

SAMPLING AND ESTIMATION

SAMPLING ERROR

Sampling error is defined as the difference between a population parameter and sample statistic. For example, the symbol μ refers to the known mean of a population whereas the symbol $\bar{x}$ refers to the arithmetic mean of a sample drawn from the population. Similarly, the symbol σ refers to the population standard deviation while the symbol s refers to the sample standard deviation. Sampling error, therefore, is the difference between a population parameter and a sample estimate of that parameter.

Quantification of the degree of sampling error, commonly referred to as standard error, can be approached either experimentally or theoretically.

Experimentally: Several samples can be drawn from a population and combined to form a distribution of sample averages. This sampling distribution, since it represents all possible variations in sampling results, represents the range of possible errors associated with sampling. Therefore, the standard deviation of the sampling distribution is the measure of standard error with sampling.

Theoretically: The sampling distribution can be approximated by computing the number of different ways the value of a population parameter can occur in different combinations. For example, for a population of eight items taken four at a time, there are 70 different ways the values can occur.

The variation of these 70 different values approximates the sampling distribution where, again, the standard deviation is the measure of standard error.

SAMPLING DISTRIBUTION FOR MEANS

The sampling distribution for means can be illustrated by drawing 100 samples of 10 each from a population of 1,611 cards ranging in number from 1 to 25.

Theoretically, the standard deviation of the sampling distribution can be calculated using the formula:

$$s_{\bar{x}} = \frac{\sigma}{\sqrt{n}}$$

where $s_{\bar{x}}$ is the standard error of the mean (the standard deviation of the sampling distribution), σ is the population standard deviation, and n is the size of the individual samples.

CARD POPULATION

CARD NUMBER	QTY.	CARD NUMBER	QTY.	CARD NUMBER	QTY.
1	1	10	127	19	43
2	2	11	154	20	26
3	4	12	174	21	14
4	7	13	181	22	7
5	14	14	174	23	4
6	26	15	154	24	2
7	43	16	127	25	1
8	67	17	96		
9	96	18	67	TOTAL:	1611

$\mu = 13$ $\sigma = 3.56$

SAMPLE AVERAGES

10.9	13.5	11.7	13.3	13.8	12.5	15.0	12.7	14.3	12.7
13.0	13.2	14.0	13.1	13.2	12.7	12.6	11.5	13.2	12.9
12.4	13.9	14.1	12.2	13.1	11.7	11.5	14.6	12.6	12.9
13.9	14.0	11.7	12.1	13.2	13.6	14.4	14.0	12.2	13.7
12.6	11.6	11.8	12.1	13.1	13.2	12.5	14.0	16.4	12.2
12.6	13.7	13.6	14.0	12.1	13.2	14.8	13.6	12.5	14.5
14.4	13.9	13.8	15.1	14.2	14.4	13.5	12.7	14.5	14.4
12.9	11.3	14.5	13.0	12.0	13.3	12.7	14.8	11.3	11.0
12.7	14.6	15.2	14.1	16.1	14.7	12.3	11.2	14.3	14.7
12.9	12.3	11.9	14.0	14.5	12.4	11.9	12.3	12.4	12.6

$\bar{\bar{X}} = 13.20$ $S_{\bar{X}} = 1.13$

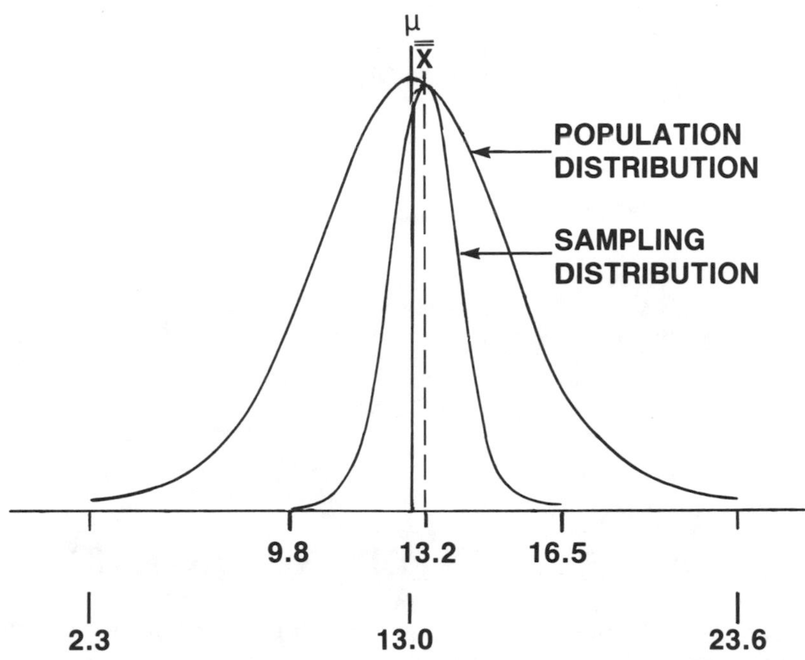

Therefore, instead of drawing 100 samples of 10 each and computing the mean and standard deviation of the sampling distribution, which was 1.13, we can simply compute the standard error associated with a sample size of 10.

$$s_{\bar{x}} = \frac{\sigma}{\sqrt{n}} = \frac{3.56}{\sqrt{10}} = 1.12$$

As the formula indicates, the dispersion of the sampling distribution is directly related to the dispersion of the population and inversely related to the size of the sample.

Dispersion: The greater the dispersion of the population, the greater the dispersion of the sampling distribution.

Sample Size: The smaller the size of the sample, the greater the dispersion of the sampling distribution. For a sample size of 1, the dispersion of the sampling distribution is equal to the dispersion of the population.

Another characteristic of the sampling distribution is the tendency toward normality regardless of the shape of the population distribution (Figure 7.1).

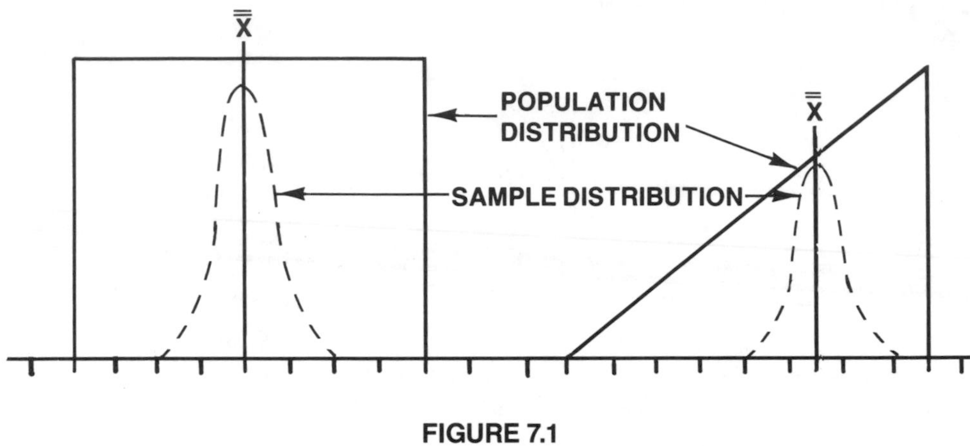

FIGURE 7.1

SAMPLING DISTRIBUTION FOR PROPORTIONS

While the sampling distribution for means will approximate the normal curve regardless of the shape of the population distribution, this is only true for proportions if the population p value is equal to or greater than 0.1 and the value of np (the product of the subgroup sample size times p) is equal to or greater than 5.

When these conditions are met, the following formula can be used to estimate the standard deviation of the sampling distribution (i.e., the standard error):

$$s_p = \sqrt{\frac{\bar{p}\,\bar{q}}{n}}$$

where s_p is the standard error of the estimate, $\bar{p}$ is the average proportion of unacceptable events, $\bar{q}$ is the average proportion of acceptable events, and n is the subgroup sample size.

When the above conditions are not met, when p is less than 0.1 and np is less than 5, the sampling distribution will be skewed to the right making sample estimates erroneous if based on the normal curve.

For individual sample estimates, if the value of p is not approximately equal to 0.1, or if the value of np is not equal to or greater than 5, estimates should be based on the binomial or poisson distributions. In these instances, the following rules apply:

Rule 1: If p is greater than 0.1 and np is less than 5, estimates should be based on the binomial distribution.

Rule 2: If p is less than 0.1, regardless of the value of np, estimates should be based on the poisson distribution.

While these rules provide general guidelines, the appropriate method is to perform a "goodness of fit" test of the sample data and, based on the results, select the distribution which best approximates the data.

CONFIDENCE LIMITS

A distinction is commonly made between point estimates and interval estimates. A point estimate, like the mean or standard deviation, is a value obtained by direct sample calculations. An interval estimate is a point estimate with confidence limits.

Note: 90% confidence limits include the area under the curve between the 5% point and the 95% point. Therefore, the total area outside the curve is 10%, 5% for the lower tail and 5% for the upper tail.

TABLE 7.1 z Values for Confidence Limits

Confidence Level (%)	Two-Sided Test ($\alpha/2$)	Z Value
99	.005	2.58
95	.025	1.96
90	.05	1.64
80	.10	1.28

CONFIDENCE LIMITS FOR MEANS OF LARGE SAMPLES

The calculation of confidence limits for the mean of large samples from a normal population (sample size equal to or greater than 30) is a relatively simple procedure in that the calculations are based on the normal curve:

$$c_L = \bar{x} \pm z \frac{s}{\sqrt{n}}$$

Where c_L is the predicted confidence limits for the population mean, $\bar{x}$ is the sample mean, z is the probability value associated with the area under the curve, s is the sample standard deviation, and n is the sample size.

Example 7.1
For a sample size of 50, a sample mean of 20, and a standard deviation of 2, what are the 90% confidence limits for the predicted population mean?

$$c_L = \bar{x} \pm z\left(\frac{s}{\sqrt{n}}\right)$$

$$= 20 \pm 1.64\left(\frac{2}{\sqrt{50}}\right)$$

$$= 20 \pm 0.46 = [19.54 \text{ to } 20.46]$$

CONFIDENCE LIMITS FOR MEANS OF SMALL SAMPLES

The method used in determining confidence limits for small samples, drawn from a normal population, is similar to the method used for large samples. With small samples, however, the t distribution is used instead of the normal distribution to establish the z value (in this case the t value). Also, opposed to the normal distribution, the t distribution changes shape in relation to the number of degrees of freedom ($n - 1$). Therefore, the value of t used in calculating the confidence limits will vary depending on the degrees of freedom.

t Distribution: The t distribution, like the normal distribution, is symmetrical about the mean. It is, however, thicker at the tails and flatter than the normal curve for small numbers of degrees of freedom. A different curve exists for each number of degrees of freedom, at least until the degrees of freedom increase to a point where the t distribution approaches the normal curve form (Figure 7.2).

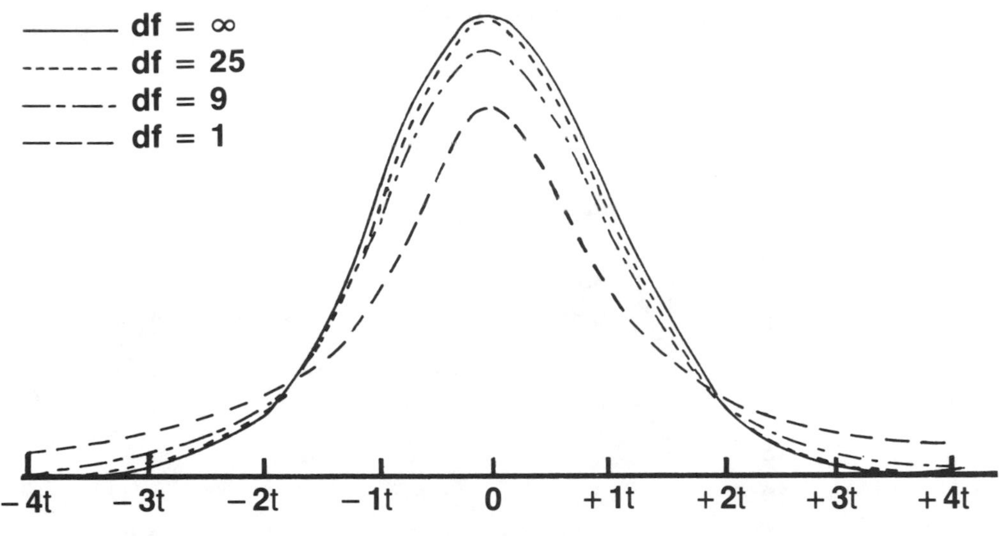

FIGURE 7.2

Degrees of Freedom: The number of degrees of freedom is the number of values that are free to vary. It is closely associated with the sample size in that the degrees of freedom for a single set of data is n – 1. For two sets of data, as in the correlation of two variables, the degrees of freedom is n – 2; n_1 – 1 for the first set of data (variable 1) plus n_2 – 1 for the second set of data (variable 2). If three sets of data are being considered, then the degrees of freedom is n – 3 (i.e., $n_1 - 1 + n_2 - 1 + n_3 - 1$).

The sum of 8 + 6 + 9 + 4 is 27. Consequently, if any three numbers are known, the remaining number is fixed.

$$8 + 6 + 9 + (x) = 27$$

Likewise, if (x) is 4, which it must be to obtain a sum total of 27, then the remaining numbers are free to vary. In this manner, 8 can become 7 if 6 also becomes 7. Nine can become 10 if 6 becomes 5, etc. Therefore, the degrees of freedom (df) for a set of four numbers is n – 1 or 3.

TABLE 7.2 t Values for Confidence Limits

Degrees Of Freedom (df)	Confidence Level			
	99%	**95%**	**90%**	**80%**
1	63.66	12.71	6.31	3.0%
2	9.93	4.30	2.92	1.88
3	5.84	3.18	2.35	1.64
4	4.60	2.78	2.13	1.53
5	4.03	2.57	2.01	1.47
10	3.17	2.23	1.81	1.37
15	2.95	2.13	1.75	1.34
20	2.85	2.09	1.72	1.32
30	2.75	2.04	1.69	1.31
120	2.62	1.98	1.65	1.29
∞	2.58	1.96	1.64	1.28

Example 7.2

For a sample size of 4 (drawn from a normal population), with a sample mean of 20, and a standard deviation of 2, what are the 90% confidence limits for the predicted population mean?

$$c_L = \overline{x} \pm t\left(\frac{s}{\sqrt{n}}\right)$$

$$= 20 \pm 2.35\left(\frac{2}{\sqrt{4}}\right)$$

$$= 20 \pm 2.35 = [17.65 \text{ to } 22.35]$$

CONFIDENCE LIMITS FOR PROPORTIONS

If the sampling distribution of a proportion is approximately normal, the z value associated with the area under the normal curve can be used as the confidence factor in computing confidence limits.

For any given sample size, however, the sampling distribution of a proportion becomes increasingly skewed as p and q depart from 0.5. In this situation, confidence limits should not be based on the normal curve.

Whether the sampling distribution can be represented by a normal distribution depends on two factors, the size of the sample and the value of p.

Confidence limits can be based on the normal distribution if the value of np is equal to or greater than 5 and if the value of p or q, whichever is smaller, is equal to or greater than 0.1. Thus, the minimum conditions where the normal curve is appropriate is when $p = 0.5$ and $n = 10$ ($np = 5$) or when $p = 0.1$ and $n = 50$ ($np = 5$) (Figure 7.3).

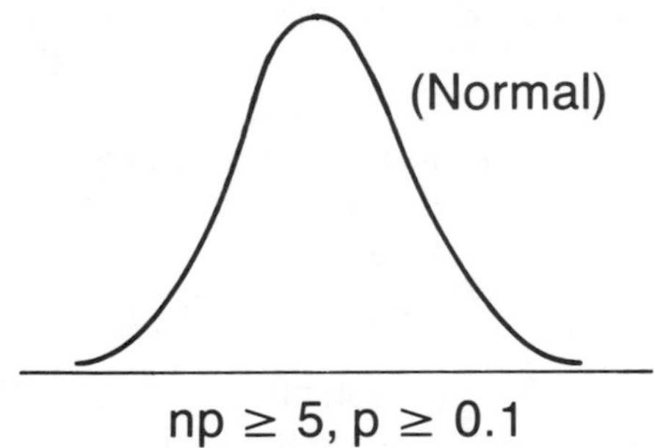

FIGURE 7.3

The normal curve is not appropriate when $p = 0.1$ and $n = 20$ ($np = 2$), or when $p = 0.04$ and $n = 100$ ($np = 4$)(Figure 7.4).

Normal Distribution: When the value of np is equal to or greater than 5, the following formula for a normal distribution can be used to compute the confidence limits for proportions:

$$c_L = p \pm z\sqrt{\frac{pq}{n}}$$

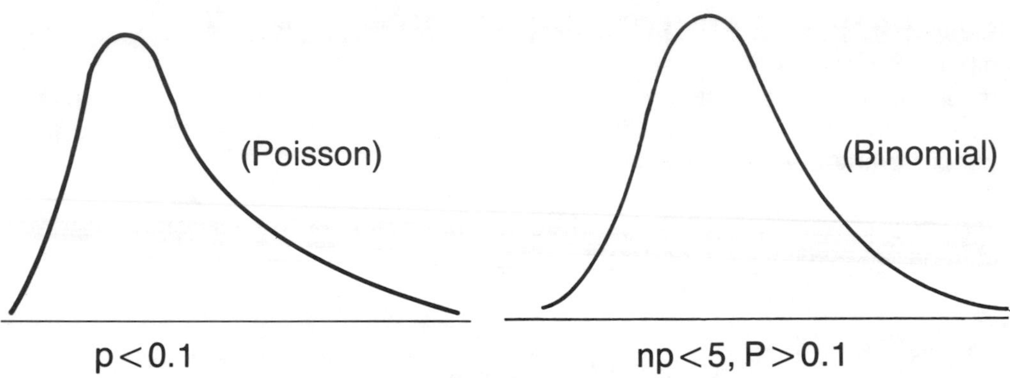

FIGURE 7.4

where c_L is the confidence limits for the predicted population proportion, p is the sample proportion of unacceptable events, z is the probability value associated with the area under the normal curve, q is the sample proportion of acceptable events, and n is the sample size.

In practical application, the value of p and the associated confidence limits are converted to percent after all calculations are performed. This conversion is accomplished by multiplying the values by 100.

Example 7.3

In a sample size of 100 drawn from a population of 1,000, 10 unacceptable events are observed. Find the 90% confidence limits for the population proportion of unacceptable events.

$$p = \frac{f}{n} = \frac{10}{100} = .10$$

$$q = 1 - p = 1 - .10 = .90$$

$$c_L = p \pm z\sqrt{\frac{pq}{n}}$$

$$= .10 \pm 1.64\sqrt{\frac{(.10)(.90)}{100}}$$

$$= .10 \pm .049 = [.051 \text{ to } .149]$$

$$= 10\% \pm 4.9\% = [5.1\% \text{ to } 14.9\%]$$

Binomial Distribution: When n is small or when np is less than 5, the binomial distribution provides the best estimate of the population distribution. However, since binomial calculations are considerably tedious to perform, curves are usually used to determine confidence limits.

Example 7.4

In a sample size of 10, 2 unacceptable events were observed. What are the 95% confidence limits for the population proportion of unacceptable events?

$$p = \frac{f}{n} = \frac{2}{10} = 0.20$$

Go to Table 7.3 and locate p = .20, then up to the first curve marked 10. Now go to the left and read that the lower confidence limit is approximately .025. Go back to p = .20 and up to the second curve marked 10. Go to the left again and read that the upper confidence limit is approximately .555.

Poisson Distribution: When the value of p is small (less than 0.1), regardless of the value of np, the poisson or "rare event" distribution provides a good approximation of the population distribution. Like the binomial distribution, calculations using the poisson distribution are tedious to perform and curves are usually used to determine confidence limits.

Example 7.5

In a sample of 500, 5 unacceptable events were observed. What are the 90% confidence limits for the population proportion of unacceptable events?

For the upper 90% confidence limit, enter the table of poisson curves from the probability of occurrence of 0.05 (Table 7.4).

Go to the right to the curve marked 5, the number of unacceptable events in the sample. Then go down and read the value of np. In this case, np = 10.5. Thus, the upper 90% confidence limit is:

$$p = \frac{np}{n} = \frac{10.5}{500} = 0.02$$

For the lower confidence limit, enter the table from the probability of occurrence of 0.95. Go to the curve marked 5; then go down and read that the value of np equal 2.6. Thus, the lower 90% confidence limit is:

$$p = \frac{np}{n} = \frac{2.6}{500} = 0.005$$

TABLE 7.3 Binomial Curves for Confidence Limits
95% Confidence Limits

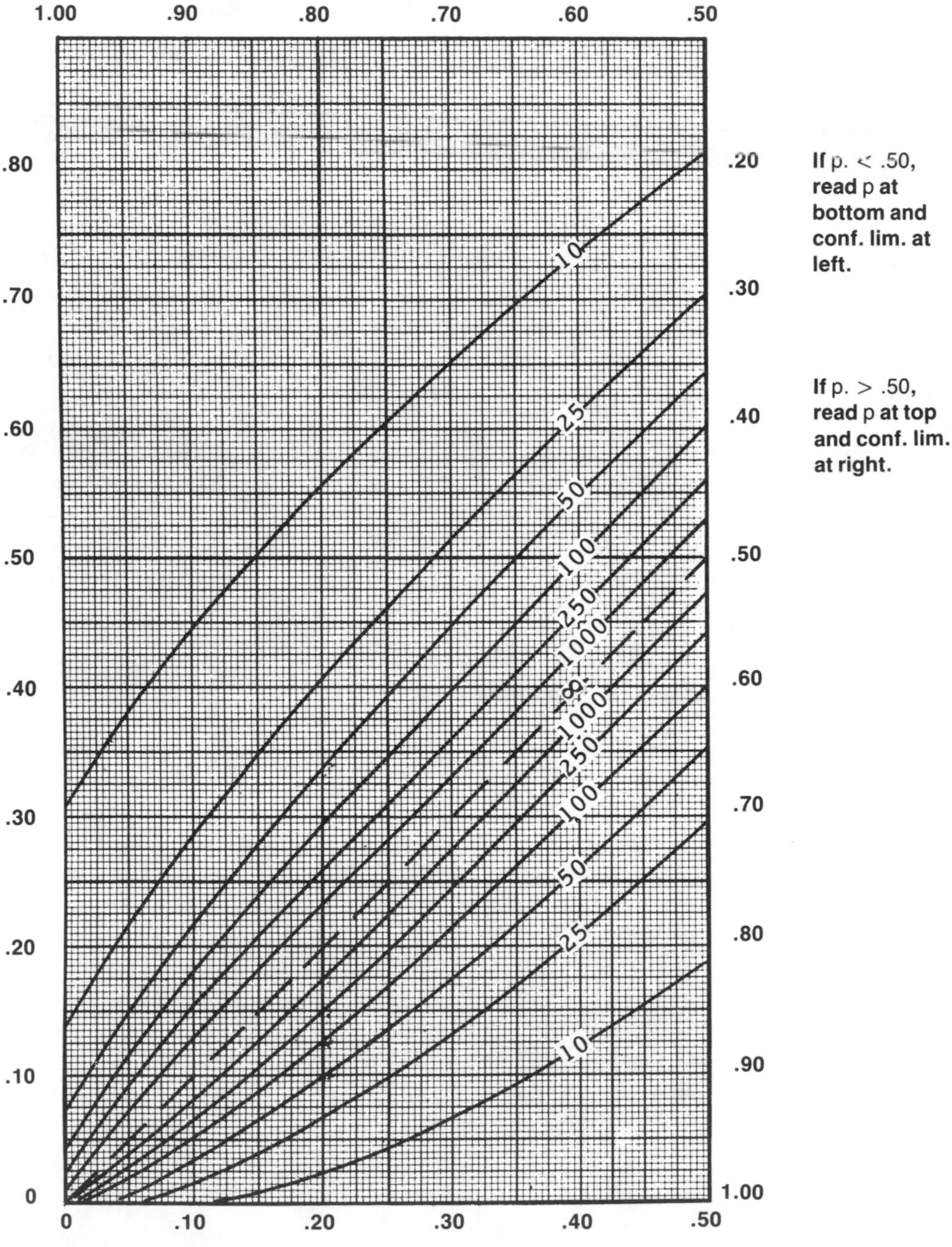

If p. < .50, read p at bottom and conf. lim. at left.

If p. > .50, read p at top and conf. lim. at right.

TABLE 7.4 Poisson Curves for Confidence Limits

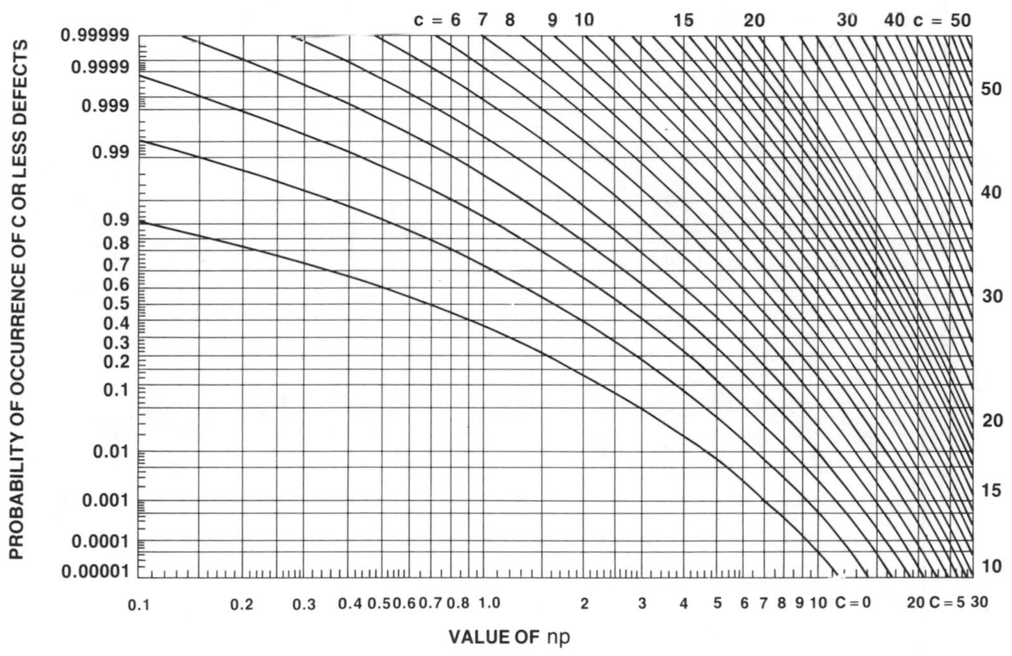

Exercise Worksheets

Exercise 7.1

CONFIDENCE LIMITS FOR MEANS OF LARGE SAMPLES
(NORMAL DISTRIBUTION)

For a sample size of 100, with a mean of 22 and a standard deviation of 3, calculate 95% confidence limits for the mean.

$$C_L = \bar{x} \pm z\left(\frac{s}{\sqrt{n}}\right)$$

$$= [\quad\quad] \pm [\quad\quad]\left(\frac{[\quad\quad]}{\sqrt{[\quad\quad]}}\right)$$

$$= [\quad\quad] \pm [\quad\quad]$$

Upper confidence limit = [ ]

Lower confidence limit = [ ]

Exercise 7.2

CONFIDENCE LIMITS FOR MEANS OF SMALL SAMPLES (t DISTRIBUTION)

For a sample size of 11, with a mean of 5 and a standard deviation of 1.2, calculate 95% confidence limits for the mean.

$$c_L = \bar{x} \pm t\left(\frac{s}{\sqrt{n}}\right)$$

$$= [\quad] \pm [\quad]\left(\frac{[\quad]}{\sqrt{[\quad]}}\right)$$

$$= [\quad] \pm [\quad]$$

Upper confidence limit = [ ]

Lower confidence limit = [ ]

Exercise 7.3

CONFIDENCE LIMITS FOR PROPORTIONS USING NORMAL DISTRIBUTION

$(p \geq .10 \; np \geq 5)$

For a sample size of 75 with 9 defectives, calculate 95% confidence limits for the population percent defective.

$$c_L = p \pm z \sqrt{\frac{(p)(q)}{n}}$$

$$= [\quad] \pm [\quad] \sqrt{\frac{[\quad] \times [\quad]}{[\quad]}}$$

Upper confidence limit = [ ]

Times 100 = [ ]

Lower confidence limit = [ ]

Times 100 = [ ]

Exercise 7.4

CONFIDENCE LIMITS FOR PROPORTIONS USING BINOMIAL DISTRIBUTION

($np < 5$)

For a sample size of 25 with 3 defectives, determine 95% confidence limits for the population percent defective.

$$p = \frac{f}{n} = [\qquad]$$

Lower confidence limit = [ ]

Times 100 = [ ]

Upper confidence limit = [ ]

Times 100 = [ ]

Exercise 7.5

CONFIDENCE LIMITS FOR PROPORTIONS USING POISSON DISTRIBUTION

$(p < 0.1)$

For a sample size of 500 with 4 defectives, determine the 90% confidence limits for population percent defective.

$$p = \frac{f}{n} = [\qquad]$$

Lower confidence limit = [ ]

Times 100 = [ ]

Upper confidence limit = [ ]

Times 100 = [ ]

Exercise 7.6

FIELD SERVICE ESTIMATE

You receive a call from the field service manager requesting that you investigate the quality level on a certain product type.

You previously predicted a 1% attention rate at the customer site and the service department staffed accordingly.

With 1,000 units already shipped to customers, you decide to put 500 units in the warehouse on hold. From this 500 you take a random 10% sample and find 5 defectives.

Based on 95% confidence, for the 1,000 units already shipped, what is your best estimate of the maximum number of units that will need attention at the customer site?

SECTION 8

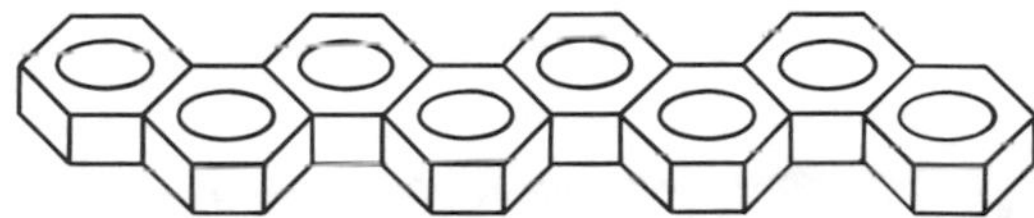

SIGNIFICANCE TESTING

TESTS OF SIGNIFICANCE

Tests of significance are performed to determine if a real difference exists between two samples — i.e., are the samples from the same population or from different populations?

Hypothesis Testing: A hypothesis is a statement that something is true. An experimental hypothesis is a prediction of how the hypothesis statement can be supported by data. Hypothesis testing, therefore, is the act of determining whether the experimental hypothesis is true or false based on the degree of difference between two samples.

If the magnitude of the difference between two samples is so great that it exceeds the probability of being attributed to sampling error alone, then the experimental hypothesis is accepted. By accepting the experimental hypothesis, the conclusion is made that the hypothesis statement is true, that a real or significant difference exists between the two sample populations.

Statement Error: There are two kinds of wrong decisions (errors) which may be made in either accepting or rejecting the experimental hypothesis. First, the experimental hypothesis may be accepted as true when it is false. This is a "Type 1 Error." Conversely, the experimental hypothesis may be rejected as false when it is true. This is a "Type 2 Error." Type 1 and Type 2 errors are represented as alpha errors ($\propto$) or beta errors (β) (Figure 8.1).

	TRUE	FALSE
ACCEPT	CORRECT DECISION	$\propto$
REJECT	β	CORRECT DECISION

FIGURE 8.1

A Type 1 or ∞ error is the error commonly associated with accepting or not accepting the experimental hypothesis. It is also common practice to use either a .05 or .01 level of significance. For example, if there is only a 5% chance (5 times in 100) of saying that there is a difference when a difference does not exist, then the difference is said to be significant at the .05 or 5% level. For the .01 level, since there is only a 1% chance (1 time out of 100) of saying that there is a difference when there is not, then the difference is said to be significant at the 1% level.

Decisions regarding the rejection of the experimental hypothesis, other than to say that the experimental hypothesis is either accepted or not accepted, is a different situation. In this case, since we're dealing with a Type 2 or β error, which increases as the sample size decreases, significance may not be indicated when a real significance exists. For this reason, and for simplicity, the preferred method is to either accept or not accept the experimental hypothesis.

DIRECTIONAL AND NON-DIRECTIONAL TESTS

A directional test is when the experimental hypothesis specifies a direction. For example, that the mean of one sample will be significantly greater than the mean of another sample (Figure 8.2). For a non-directional test, since a direction is not proposed, the hypothesis is accepted if the mean of one sample is either significantly greater or less than another sample (Figure 8.3).

The importance of specifying a directional or non-directional test is that it defines, for a given level of significance, the critical value of probability (i.e., for a directional test at a .05 level of significance, the critical value of z is 1.64). This is because only one tail of the distribution is used. For a non-directional test at a .05 level of significance, since both tails are now used, the critical value is 1.96; 2.5% of the area under the curve to the right of +1.96 standard deviation above the mean and 2.5% to the left of −1.96 standard deviation below the mean. The area outside these limits, therefore, is 5% of the total area under the curve. When an occurrence happens in this area, since it can only happen 5% of the time due to chance alone (5 times in 100), the difference is said to be significant at the .05 level.

TESTS OF SIGNIFICANCE FOR MEANS: t DISTRIBUTION

Significance tests for means are classified as either tests for independent samples or test for correlated samples.

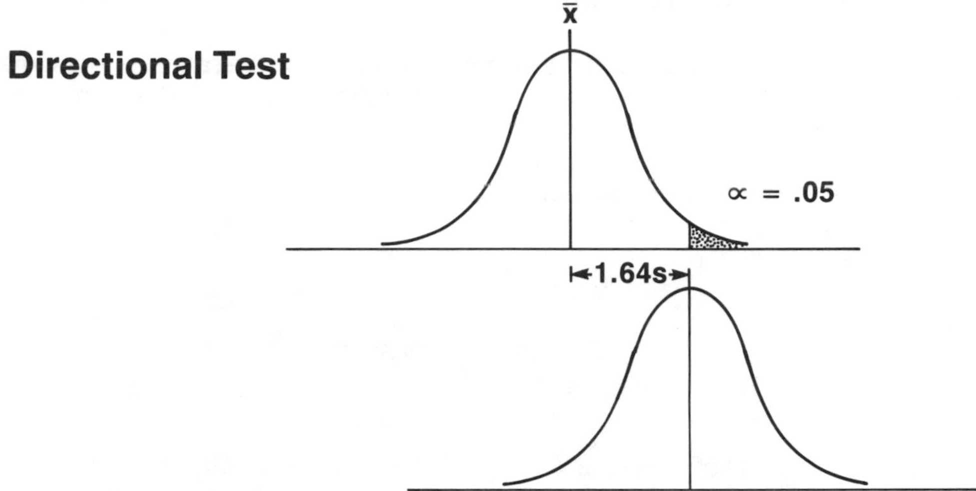

FIGURE 8.2

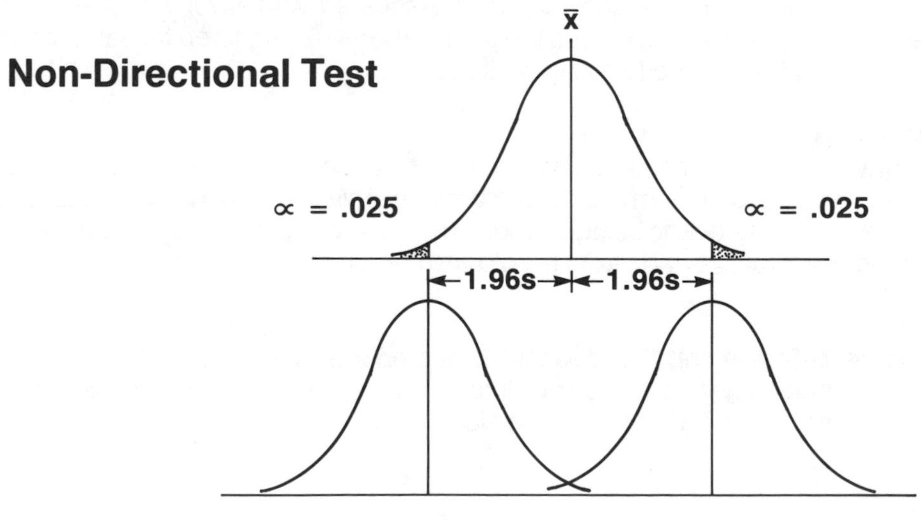

FIGURE 8.3

Test for Independent Samples

Significance tests for independent samples are used to determine if the scores between two different samples, drawn from two different populations, differ more than would be expected due to chance alone.

This test assumes that the population distributions are normal and have equal variances (standard deviation). When this is true, the following formula can be used to calculate the value of t:

$$t = \frac{\overline{x}_1 - \overline{x}_2}{\sqrt{\frac{(s^2_{1,2})}{n_1} + \frac{(s^2_{1,2})}{n_2}}}$$

where $\overline{x}_1$ is the mean of sample 1, $\overline{x}_2$ is the mean of sample 2, $(s^2_{1,2})$ is the combined or pooled sample variance, and n_1 and n_2 are the sample sizes for samples 1 and 2.

The combined sample variance $(s^2_{1,2})$ is computed using the following formula:

$$(s^2_{1,2}) = \frac{\left[\Sigma x_1^2 - \frac{(\Sigma x_1)^2}{n_1}\right] + \left[\Sigma x_2^2 - \frac{(\Sigma x_2)^2}{n_2}\right]}{n_1 + n_2 - 2}$$

where Σx_1^2 and Σx_2^2 are the sum of the squared values for samples 1 and 2 respectively, $(\Sigma x_1)^2$ and $(\Sigma x_2)^2$ are the sum of the values squared for samples 1 and 2, and n_1 and n_2 are the two sample sizes.

Example 8.1

A newly developed process is about to be introduced into manufacturing. To a .05 level of significance, determine if this new process significantly increases the average value of the characteristic being considered. For the purpose of this example, assume that the population of parts from the old and new process are normal and have equal variance.

Hypothesis Statement: The mean value of products produced by the new process will be significantly greater than products produced by the old process. **Note:** Since a direction is specified, this is a one-sided test.

MEASURED VALUES			
NEW PROCESS (SAMPLE 1)		OLD PROCESS (SAMPLE 2)	
VALUE OF x	x^2	VALUE OF x	x^2
90	8100	87	7569
88	7744	89	7921
93	8649	91	8281
94	8836	92	8464
88	7744	86	7396
87	7569	84	7056
89	7921	90	8100
92	8464	89	7921
93	8649	87	7569
85	7225	88	7744
$\Sigma x = 899$	$\Sigma x^2 = 80{,}901$	$\Sigma x = 883$	$\Sigma x^2 = 78{,}021$
$(\Sigma x)^2 = 808{,}201$	$\bar{x} = 89.9$	$(\Sigma x)^2 = 779{,}689$	$\bar{x} = 88.3$
$n_1 = 10$		$n_2 = 10$	

Step 1: Compute the combined sample variance.

$$(s^2_{1,2}) = \frac{\left[\Sigma x_1^2 - \frac{(\Sigma x_1)^2}{n_1}\right] + \left[\Sigma x_2^2 - \frac{(\Sigma x_2)^2}{n_2}\right]}{n_1 + n_2 - 2}$$

$$= \frac{\left[80{,}901 - \frac{808{,}201}{10}\right] + \left[78{,}021 - \frac{779{,}689}{10}\right]}{10 + 10 - 2}$$

$$= \frac{80.9 + 52.1}{18} = [7.4]$$

Step 2: Compute the value of t.

$$t = \frac{\bar{X}_1 - \bar{X}_2}{\sqrt{\frac{(s^2{}_{1,\,2})}{n_1} + \frac{(s^2{}_{1,\,2})}{n_2}}}$$

$$= \frac{89.9 - 88.3}{\sqrt{\frac{7.4}{10} + \frac{7.4}{10}}}$$

$$= \frac{1.6}{1.2} = [1.3]$$

To determine the critical value of t (see Table 8.2, page 202), read down the degrees of freedom (df) column to 18 (i.e., $n_1 - 1 + n_2 - 1 = 18$). Now go across to the column labeled .05 for a one-sided test. Read that the critical value of t is 1.734. Therefore, since the calculated value of t (1.3) does not equal or exceed the critical value of t (1.734), the experimental hypothesis is not accepted. In other words, the new process is not significantly better than the old process.

Tests for Correlated Samples

Significance tests for correlated samples are used when a single sample is studied under two separate conditions (i.e., a before-after comparison). This test, like the test for independent samples, assumes a normal population distribution and an equal before-after variance.

The correlated sample method is commonly known as the difference method. In this case, the sum of the differences between the first and second observation ($x_1 - x_2$) is used to compute the value of t.

$$t = \frac{\Sigma D}{\sqrt{\frac{(n)(\Sigma D^2) - (\Sigma D)^2}{n - 1}}}$$

Example 8.2
On a given product, to a .05 level of significance, determine if extreme environmental exposure affects the output voltage.

Hypothesis Statement: Extreme environmental exposure will significantly affect output voltage. **Note:** Since a direction is not specified, this is a two-sided test.

SAMPLE NUMBER	MEASURED VALUE BEFORE X_1	MEASURED VALUE AFTER X_2	D	D^2
1	7	5	2	4
2	9	15	−6	36
3	4	7	−3	9
4	15	11	4	16
5	6	4	2	4
6	3	7	−4	16
7	9	8	1	1
8	5	10	−5	25
9	6	6	0	0
10	12	16	−4	16
SUM	76	89	−13	127
MEAN	7.6	8.9		

$$t = \frac{\Sigma D}{\sqrt{\dfrac{(n)(\Sigma D^2) - (\Sigma D)^2}{n - 1}}} = \frac{-13}{\sqrt{\dfrac{(10)(127) - (-13)^2}{9}}} = [-1.17]$$

For a two-sided test at a .05 level of significance, with 9 degrees of freedom, the critical value of t is 2.262. Therefore, since the calculated value of t is well below the critical value of t, we cannot justifiably argue that environmental exposure affects output voltage.

Note: In this example, 9 degrees of freedom are used instead of 18 (i.e., as used for two independent samples) because the difference between the two measures are used instead of each individual measure. Thus, since there are 10 differences, the degrees of freedom is 10−1, or 9.

Tests for Means with Unequal Population Variance

The t test for independent samples assumes that each population variance is approximately the same. When this is not the case, the following formula is used to calculate the value of t:

$$t = \frac{\overline{X}_1 - \overline{X}_2}{s_{\bar{x}_1 - \bar{x}_2}}$$

where $s_{\bar{x}_1 - \bar{x}_2}$ (the standard error of the difference between the sample means) is calculated using the formula:

$$s_{\bar{x}_1 - \bar{x}_2} = \sqrt{s_{\bar{x}_1}{}^2 + s_{\bar{x}_2}{}^2}$$

where $s_{\bar{x}_1}{}^2$ and $s_{\bar{x}_2}{}^2$ (the individual sample variances) are calculated using the formula:

$$s_{\bar{x}}^2 = \frac{\Sigma (x - \overline{x})^2}{n (n - 1)}$$

Then, since the population variances are unequal, an adjusted value of t must be computed. This is accomplished by multiplying together the sample variance and critical value of t for sample 1; multiplying together the sample variance and critical value of t for sample 2; adding these two results together and dividing that result by the standard error of the difference between means. This is expressed algebraically as:

$$t_a = \sqrt{\frac{(s_{\bar{x}_1}{}^2)(t_1) + (s_{\bar{x}_2}{}^2)(t_2)}{s_{\bar{x}_1}{}^2 + s_{\bar{x}_2}{}^2}}$$

where t_a is the adjusted value of t, t_1 is the critical value of t for sample 1, and t_2 is the critical value of t for sample 2.

Example 8.3

Supplier A and B produce the same type of product. To a .05 level of significance, determine if the products from supplier B have a higher mean value than the products from supplier A.

Hypothesis Statement: Products produced by supplier B are significantly higher in measured value than products produced by supplier A (a one-sided test).

Supplier B	Supplier A
$n_1 = 13$	$n_2 = 9$
$\bar{x}_1 = 26.99$	$\bar{x}_2 = 15.10$
$\Sigma (x - \bar{x}_1)^2 = 1{,}128$	$\Sigma (x - \bar{x}_2)^2 = 1{,}269$

Step 1: Calculate the sample variance ($s_{\bar{x}}^2$) for each sample.

$$s_{\bar{x}_1}^{\,2} = \frac{\Sigma (x - \bar{x}_1)^2}{n_1 (n_1 - 1)} = \frac{1{,}128}{13(12)} = [7.23]$$

$$s_{\bar{x}_2}^{\,2} = \frac{\Sigma (x - \bar{x}_2)^2}{n_2 (n_2 - 1)} = \frac{1{,}269}{9(8)} = [17.62]$$

Step 2: Calculate the standard error of the difference between means: ($s_{\bar{x}_1 - \bar{x}_2}$).

$$s_{\bar{x}_1 - \bar{x}_2} = \sqrt{s_{\bar{x}_1}^{\,2} + s_{\bar{x}_2}^{\,2}} = \sqrt{7.23 + 17.62} = [4.98]$$

Step 3: Calculate the value of t.

$$t = \frac{\bar{x}_1 - \bar{x}_2}{s_{\bar{x}_1 - \bar{x}_2}} = \frac{26.99 - 15.10}{4.98} = [2.38]$$

Step 4: Determine critical values of t for sample 1 and 2. **Note:** For a one-sided test at a .05 level of significance, the critical values of t, as read from the t table, are 1.782 for sample 1 (12df) and 1.860 for sample 2 (8df).

Step 5: Calculate the adjusted critical value of t.

$$t_a = \sqrt{\frac{(s_{\bar{x}_1}^{\,2})(t_1) + (s_{\bar{x}_2}^{\,2})(t_2)}{s_{\bar{x}_1}^{\,2} + s_{\bar{x}_2}^{\,2}}}$$

$$= \sqrt{\frac{(7.23)(1.782) + (17.62)(1.860)}{7.23 + 17.62}} = [1.35]$$

Step 6: Compare the calculated value of t with the adjusted critical value of t to determine significance.

In this example, since the calculated value of t (2.38) is greater than the adjusted critical value of t (1.35), the conclusion is that the difference between means is significant at the 5% level. Hence, the average value of products produced by supplier B are significantly higher than products produced by supplier A.

TESTS OF SIGNIFICANCE FOR PROPORTIONS: z DISTRIBUTION

Significance tests for proportions, like the tests for means, are classified as either independent two sample tests, or before-after single sample tests.

Test for Independent Samples

Certain situations may arise where significance testing of attribute data is required. In this case, significance testing is performed to determine if two independent samples can be regarded as random samples drawn from the same population (non-significance) or random samples drawn from two separate populations (significance).

This test is applicable only if the populations are reasonably normal. A quick check of normality can be determined by multiplying the smaller value of p or q by the smaller value of n. If the result is equal to or greater than 5, then a normal distribution can be assumed and the value of z used to determine significance.

To test for significance between two sample proportions, the difference between the proportion of unacceptable events in sample 1 and the proportion of unacceptable events in sample 2, is divided by the standard error of the difference.

$$z = \frac{p_1 - p_2}{s_{p_1 - p_2}}$$

where $s_{p_1 - p_2}$ is the standard error of the difference between two proportions (based on two independent samples). The formula for computing the standard error is:

$$s_{p_1 - p_2} = \sqrt{p_{1,2}\, q \left(\frac{1}{n_1} + \frac{1}{n_2}\right)}$$

where $p_{1,2}$ (the combined proportion of unacceptable events) is obtained by combining the number of unacceptable events in each sample and dividing by the total number of samples. Now q is taken as $1 - p_{1,2}$

$$p_{1,2} = \frac{f_1 + f_2}{n_1 + n_2}$$

where f is the number of unacceptable events.

Example 8.4

Inspection data indicates a difference in the number of defects produced by shifts 1 and 2. To a .05 level of significance, determine if there is a difference between the two shifts.

Hypothesis Statement: Products produced on shift 1 have significantly fewer defects than produced on shift 2.

Shift 1	Shift 2
$n_1 = 300$	$n_2 = 500$
$f_1 = 176$	$f_2 = 384$
$p_1 = .587$	$p_2 = .768$

Step 1: Calculate the combined proportion of defects for shifts 1 and 2.

$$p_{1,2} = \frac{f_1 + f_2}{n_1 + n_2} = \frac{176 + 384}{300 + 500} = [.70]$$

Step 2: Calculate the standard error of the difference between the two proportions.

$$s_{p_1 - p_2} = \sqrt{p_{1,2}\,q\left(\frac{1}{n_1} + \frac{1}{n_2}\right)} = \sqrt{.70 \times .30\left(\frac{1}{300} + \frac{1}{500}\right)}$$

$$= [.033]$$

Step 3: Calculate the z value.

$$z = \frac{p_1 - p_2}{s_{p_1 - p_2}} = \frac{.587 - .768}{.033} = [-5.48]$$

Step 4: Consult z table (Table 8.1) to determine if the calculated z value is significant at the .05 level.

In this case, for a one-sided test at a .05 level of significance, the critical value of z is 1.64. Thus, for a calculated z value of 5.48 (+ and − signs are ignored for significance testing) we can safely state that the first shift produces significantly fewer defects.

An alternative method exists for testing the difference between independent samples. This method uses the chi square (χ^2) distribution and will be discussed later.

TABLE 8.1 z Values for Significance Testing

		Z VALUE	
SIGNIFICANCE	P_a	ONE-SIDED	TWO-SIDED
	.001	3.09	3.29
YES	.005	2.58	2.81
	.01	2.33	2.58
?	.02	2.05	2.33
	.05	1.64	1.96
	.10	1.28	1.64
	.20	0.84	1.28
NO	.30	0.52	1.04
	.40	0.25	0.84
	.50	0.00	0.67

Test for Correlated Samples

Tests for correlated proportions, like the test for correlated means, are used when single or matched samples are studied under two separate conditions. This test also assumes a normal population distribution (np ≥ 5) and equal before-after variances.

To apply a test of significance to the difference between proportions, it is helpful to form the data into a 2×2 (before-after/pass-fail) table (Figure 8.4).

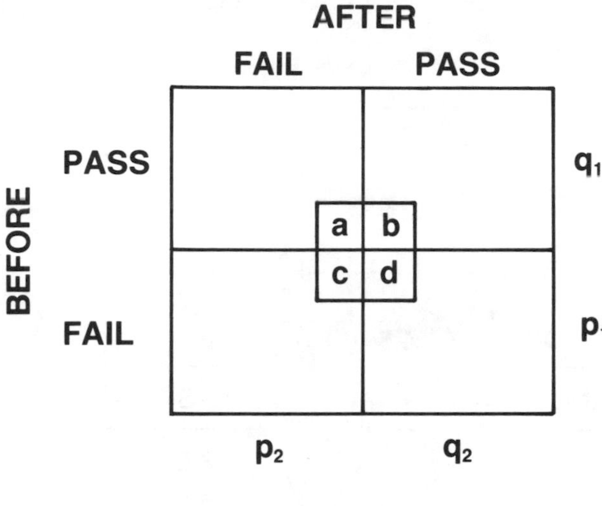

FIGURE 8.4

In this case, cells a, b, c, and d represent the proportions obtained by dividing the number of pass or fail items by the total number tested. Significance testing is then performed on the proportion passing the two tests (i.e., q_1 and q_2). The value of z is calculated using the formula:

$$z = \frac{q_1 - q_2}{\sqrt{\frac{a + d}{n}}}$$

Example 8.5

Two hundred products are tested on machine 1 and then tested on machine 2. To a .01 level of significance, determine if there is a difference between the observed test results.

CONDITION	QUANTITY	PROPORTION
PASS MACH. 1 / FAIL MACH. 2	30	.15
PASS MACH. 1 / PASS MACH. 2	110	.55
FAIL MACH. 1 / FAIL MACH. 2	50	.25
FAIL MACH. 1 / PASS MACH. 2	10	.05
TOTAL:	200	1.00

Step 1: Construct a 2×2 table and enter the appropriate proportions.

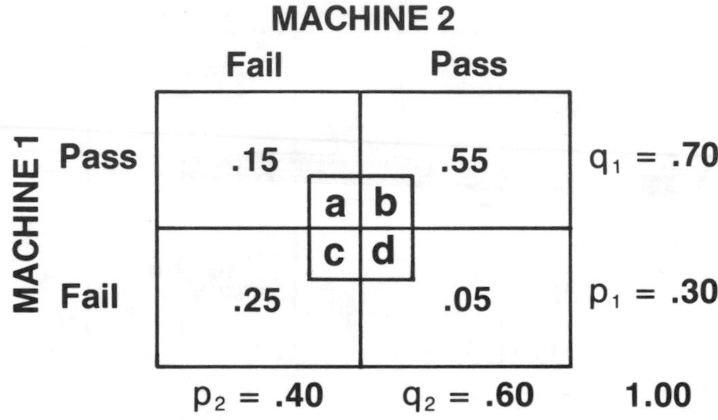

Step 2: Calculate the z value.

$$z = \frac{q_1 - q_2}{\sqrt{\dfrac{a + d}{n}}} = \frac{.7 - .6}{\sqrt{\dfrac{.15 + .05}{200}}} = [3.16]$$

In this case, the difference is significant. It exceeds the value of 2.58 required for significance at the 1% level for a two-sided test.

TESTS OF SIGNIFICANCE FOR FREQUENCIES: χ^2 DISTRIBUTION

In the preceding tests of significance for proportions, the procedures involved dividing the difference between two proportions by the standard error of the difference. This resulted in a deviation value (z) which could be referred to a table of areas (probabilities).

Where a normal distribution can be assumed, the chi square (χ^2) distribution provides an alternate, but equivalent, procedure for testing the significance of the difference between porportions. In this case, however, the quantity or frequency of events is used instead of the proportion (i.e., the quantities that pass or fail are used instead of q and p).

Tests for Independent Samples

As performed for correlated sample proportions, data for the chi square test of significance are constructed in the form of a 2×2 table. The following formula is then used to compute the value of χ^2:

$$\chi^2 = \frac{N\,[(ad) - (bc)]^2}{(a + b)\,(c + d)\,(a + c)\,(b + d)}$$

where N is the combined sample size.

Example 8.6
An outgoing quality audit of products produced in January revealed a high number of defective items. Corrective actions were immediately instituted. In March, another quality audit was performed to assess the effect of the corrective actions. To a .05 level of significance, determine if those actions were successful.

Hypothesis Statement: The number of defective items in the March audit are significantly fewer than observed in the January audit.

JAN. AUDIT	MARCH AUDIT
n_1 = 140	n_2 = 60
QTY. ACC. = 70	QTY. ACC. = 40
QTY. REJ. = 70	QTY. REJ. = 20

Step 1: Construct a 2×2 table and enter the appropriate values.

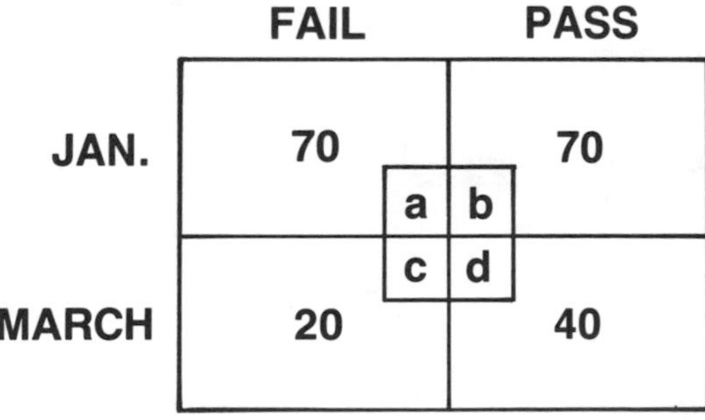

Step 2: Calculate the value of χ^2.

$$\chi^2 = \frac{N[(ad) - (bc)]^2}{(a + b)(c + d)(a + c)(b + d)}$$

$$= \frac{200[(70 \times 40) - (70 \times 20)]^2}{(70 + 70)(20 + 40)(70 + 20)(70 + 40)} = [4.71]$$

Step 3: Determine the critical value of χ^2 for a one-sided test, with 1 degree of freedom and a significance level of .05.

Note: The critical value of χ^2, as shown in Table 8.3, page 203 is based on the total area outside the χ^2 distribution and, as such, is equivalent to the total area outside the normal curve (i.e., the total area for a two-sided test of significance). For example, when the distribution is normal, χ^2 is equal to z^2. Thus the square root of χ^2 is equal to z (i.e., $\sqrt{\chi^2} = z$). For a χ^2 value of 3.84, which is the value given in the chi square table for 1 degree of freedom at a .05 level of significance, the equivalent value is $\sqrt{3.84}$ or 1.96, which is equivalent to a two-sided z test at a .05 level of significance.

In the previous example, however, we wish to determine the critical value for a one-sided test at a .05 level of significance. Therefore, we would use the critical value of chi square for a .10 level of significance (i.e., $\sqrt{2.71}$ equals a z value of 1.64), which is the critical value of χ^2 for a one-sided test based on the normal distribution.

A more straightforward method of determining the critical value of χ^2 for a one-sided test is to square the critical value of z. In this manner, the critical value of χ^2 for the above example is $(1.64)^2$ or, allowing for rounding error, 2.68.

Since the calculated value of χ^2 (4.71) far exceeds the critical value of 2.68, we can safely accept the hypothesis statement.

Tests for Correlated Samples

Where data are composed of correlated paired observations, the following χ^2 test of frequencies can be used:

$$\chi^2 = \frac{(d - a)^2}{a + d}$$

where d and a are cell frequencies in the bottom right and top left cells of a 2×2 table.

Example 8.7
Using the same example used for testing correlated sample proportions (a z test), we have: 200 products are tested on machine 1 and then tested on machine 2. To a .01 level of significance, determine if a difference exists between the two test results.

CONDITION	QUANTITY
PASS MACH. 1 / FAIL MACH. 2	30
PASS MACH. 1 / PASS MACH. 2	110
FAIL MACH. 1 / FAIL MACH. 2	50
FAIL MACH. 1 / PASS MACH. 2	10

Step 1: Construct a 2×2 table and enter the appropriate values.

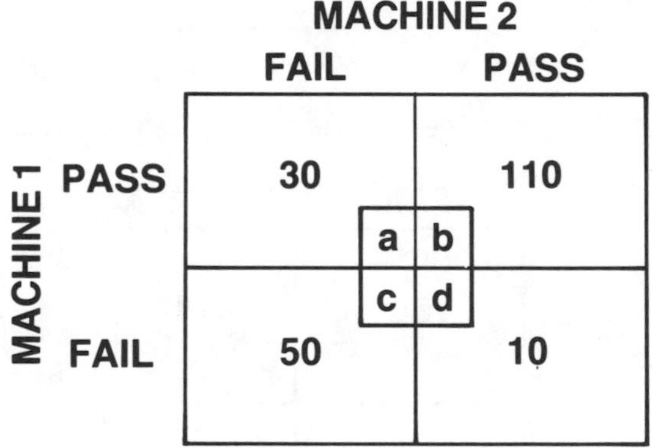

Step 2: Calculate the χ^2 value.

$$\chi^2 = \frac{(d - a)^2}{a + d} = \frac{(10 - 30)^2}{40} = [10]$$

Step 3: Refer to chi square Table 8.3 for the critical value of χ^2. For 1 degree of freedom at a .01 level of significance, the critical value of χ^2 is 6.64. Since this is a two-sided test, this value is appropriate without conversion.

Since the calculated value of χ^2 exceeds the critical value, it is safe to state that there is a significant difference between the two test machines.

Test for Same-Sample Frequencies

In the preceding example for correlated sample frequencies, suppose we simply wanted to determine if a difference existed between the two test machines without tracking the individual items through machine 1 and 2.

In this case, the data are recorded as shown below and then entered in a 2×2 table.

MACHINE NUMBER	QUANTITY		
	TESTED	PASS	FAIL
1	200	140	60
2	200	120	80
TOTAL	400		

Step 1: Construct a 2×2 table and enter the appropriate values.

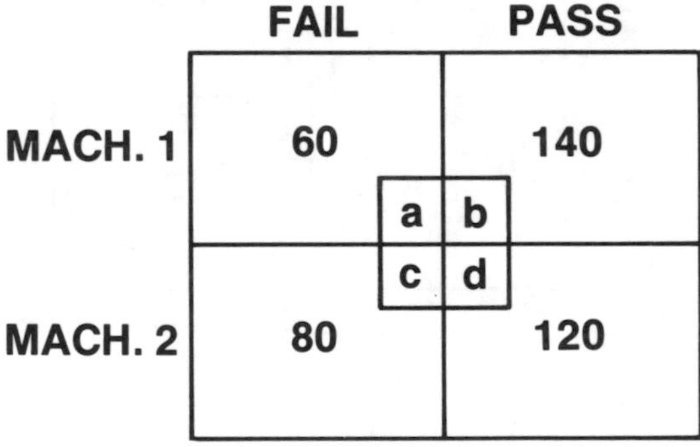

Step 2: Calculate the value of χ^2.

$$\chi^2 = \frac{N\,[(a)\,(d) - (b)\,(c)\,]^2}{(a + b)\,(c + d)\,(a + c)\,(b + d)}$$

$$= \frac{400\,[\,(60)\,(120) - (140)\,(80)\,]^2}{(60 + 140)\,(80 + 120)\,(60 + 80)\,(140 + 120)}$$

$$= [4.39]$$

Again, referring to Table 8.3, Critical Values of Chi Square, (page 203) for a two-sided test with 1 degree of freedom and a significance level of .01, we see that the critical value of χ^2 (6.64) is greater than the calculated value (4.39). Therefore, as opposed to the test for correlated sample frequencies, there is not a significant difference between the two test machines.

TEST OF SIGNIFICANCE FOR VARIANCE: F DISTRIBUTION

Occasions may arise where a test of the difference between two variances is required. In this case, the following method which assumes equality in the populations can be used to test the difference between two sample variances.

Tests for Independent Samples

Tests for independent sample variances are based on the F distribution which, like the t distribution, changes for different degrees of freedom.

To test the significance of the difference between two variances, calculate the ratio of the two unbiased sample variance estimates. These are calculated using the formula:

$$s_1^2 = \frac{\Sigma (x - \bar{x}_1)^2}{n_1 - 1}$$

$$s_2^2 = \frac{\Sigma (x - \bar{x}_2)^2}{n_2 - 1}$$

where $\Sigma (x - \bar{x})^2$ is the sum of squares of deviations.

The F ratio is then obtained by dividing one of the unbiased sample variance estimates by the other, s_1^2/s_2^2 or s_2^2/s_1^2.

No prior rules exist for deciding which variance estimate should be divided by the other. In practice, however, the larger of the two variance estimates is always divided by the smaller. In this manner, the F ratio is always greater than unity, lending itself to use of the F table.

Example 8.8
A sample of 31 products is drawn from process A and a sample of 26 from process B. To a .05 level of significance, determine if a difference exists in product variance between process A and process B.

Hypothesis Statement: Product variance resulting from process A is significantly greater than the variance of products from process B.

Process A	Process B
$n_1 = 31$ $\Sigma (x - \bar{x}_1)^2 = 1{,}926$	$n_2 = 26$ $\Sigma (x - \bar{x}_2)^2 = 2{,}875$

Step 1: Calculate the unbiased sample variance estimate for both samples.

$$s_1{}^2 = \frac{\Sigma(x - \bar{x}_1)^2}{n_1 - 1} = \frac{1{,}926}{30} = [64.2]$$

$$s_2{}^2 = \frac{\Sigma(x - \bar{x}_2)^2}{n_2 - 1} = \frac{2{,}875}{25} = [115.0]$$

Step 2: Calculate the F ratio.

$$F = \frac{s_2{}^2}{s_1{}^2} = \frac{115}{64.2} = [1.79]$$

Step 3: Determine the degrees of freedom for the greater variance estimate (listed on the top of F Table 8.4, page 204, as degrees of freedom for greater variance estimate), and for the lesser variance estimate (listed on the side of the F table as degrees of freedom for lesser variance estimate).

For example, process B has the greater variance estimate (115). Therefore, since the sample size for process B is 26, the degrees of freedom (n − 1) for the greater variance estimate is 25. Accordingly, for the lesser variance estimate, the degrees of freedom is 30.

Since the calculated value of F (1.79) is less than the critical value of F (1.89), the hypothesis statement is not accepted. Thus, there is not a significant difference in variance between process A and B.

CRITICAL VALUES OF F

df Lesser Variance Estimate	5 Percent Level of Significance							
	df Greater Variance Estimate							
		12	15	20	24	30	40	
25								
26								
27								
28								
29								
30					1.89			
40								
60								
120								

Note: Since this is a one-sided test, the critical value of F is read directly from F Table 8.4. For a two-sided test, the critical value of F is simply doubled.

Tests for Correlated Samples

Correlated variance tests are performed to determine if a significant change occurs when identical samples are subjected to two different conditions.

This test accounts for the correlation between the before-after condition, and therefore, the correlation coefficient (r) must be calculated. By using the correlation coefficient in the following formula, the t distribution, or critical value of t, can be used to determine significance:

$$t = \frac{(s_1^2 - s_2^2)\sqrt{n - 2}}{\sqrt{(4)(s_1^2)(s_2^2)(1 - r^2)}}$$

where s_1^2 and s_2^2 are the unbiased sample variance estimates and r is the correlation coefficient.

Example 8.9
Thirty-eight samples are measured for hardness before and after heat treatment. To a .05 level of significance, determine if a difference in variance exists between the before-after measures.

Hypothesis Statement: The effect of heat treatment will result in a significant decrease in sample variance.

Step 1: Use before-after measurements to calculate the correlation coefficient. For this example assume a value of .60.

Step 2: Calculate the value of s_1^2 and s_2^2.

$$s_1^2 = \frac{\Sigma(x - \bar{x}_1)^2}{n_1 - 1}$$

$$s_2^2 = \frac{\Sigma(x - \bar{x}_2)^2}{n_2 - 1}$$

For the purpose of this example, assume a s_1^2 value of 153.20 and a s_2^2 value of 102.51.

Step 3: Calculate the value of t.

$$t = \frac{(s_1^2 - s_2^2)\sqrt{n - 2}}{\sqrt{(4)(s_1^2)(s_2^2)(1 - r^2)}} = \frac{(153.20 - 102.51)\sqrt{38 - 2}}{\sqrt{(4)(153.20)(102.51)(1 - .36)}}$$

$$= [1.52]$$

For a one-sided test with 36 degrees of freedom (38 − 2), a t value equal to or greater than 1.69 is required for significance at a 5% level. Therefore, the hypothesis statement is not accepted.

DETERMINING SAMPLE SIZES

Since there are no pre-established minimum or maximum limits for the size of an acceptable sample, the usual criteria involves three major considerations: the importance of the study, the amount of time and money available, and the amount of error that can be tolerated.

Sample Size for Means

Tests of significance are based on the standard error associated with the difference between two means. Therefore, since the magnitude of the standard error is determined by the size of the sample, the sample must be sufficiently large enough to detect significance when significance exists.

The formula for determining the minimum samples size is:

$$n = \frac{(s)^2 \times (2z)^2}{(x_d - \bar{x})^2}$$

where s and $\bar{x}$ are computed sample statistics (standard deviation and mean), z is the number of standard deviation units from the mean for a given level of significance, and x_d is the displacement between means.

To compute the sample statistics ($\bar{x}$ and s), a sample size of at least 20 is required. Then, if the calculated value of n is 20 or less, no further samples are required. If the required n value is greater than 20, additional samples should be drawn and $\bar{x}$ recalculated.

Example 8.10
Given the problem to determine the sample size required to detect a significant difference for a displacement of one-half standard deviations between two means, for a two-sided test at a .05 level, the procedure is:

Step 1: Calculate the mean and standard deviation for a sample size of 20. For illustration, assume a mean of 15 and a standard deviation of 3.

Step 2: Determine the value of z for a two-sided test at a .05 level of significance. This value is 1.96.

Step 3: Determine the value of x_d (i.e., the displacement from $\bar{x}$ to be detected). For this example, the displacement is given as one-half standard deviation. Therefore, for a sample standard deviation of 3 and a mean of 15, the value of x_d is 15 plus 1.5, or 16.5.

Step 4: Given the stated values ($\bar{x} = 15, x_d = 16.5, s = 3$, and $z = 1.96$), compute the required sample quantity.

$$n = \frac{(s)^2 \times (2z)^2}{(x_d - \bar{x})^2} = \frac{(3)^2 \times (2\times 1.96)^2}{(16.5 - 15)^2}$$

$$= 61.46 = [62]$$

Given the same conditions for a one-sided test, the required sample size is:

$$n = \frac{(s)^2 \times (2z)^2}{(x_d - \bar{x})^2} = \frac{(3)^2 \times (2 \times 1.64)^2}{(16.5 - 15)^2} = [43]$$

Given the same condition for a one-sided test with a displacement of 1 standard deviation between x_d and $\bar{x}$, the required sample size is:

$$n = \frac{(s)^2 \times (2z)^2}{(x_d - \bar{x})^2} = \frac{(3)^2 \times (2 \times 1.64)^2}{(18 - 15)^2} = 10.7 = [11]$$

Sample Size for Proportions

Since attribute information is less sensitive than actual measure data, the sample size required to determine significance is greatly higher. For example, a sample size of at least 100 is needed to detect a proportion defective of 1%.

The formula for determining the minimum sample size for proportions is:

$$n = \frac{(s_p)\,(z)^2}{(p_d - p)^2}$$

where s_p and p are computed sample statistics (standard deviation and sample fraction defective), z is the number of standard deviation units for a given level of significance, and $p_d - p$ is the fraction defective displacement to be detected.

If previous process data are not available, a sample of at least 100 items should be drawn and the sample fraction defective (p) and standard deviation (s_p) calculated.

$$p = \frac{f}{n} \text{ and } s_p = \sqrt{\frac{pq}{n}}$$

Example 8.11

Given the problem to determine the sample size required to detect a significant difference for a displacement of 1 standard deviation between two sample proportions, for a two-sided test at a .05 level, the procedure is:

Step 1: Draw a sample of 100 items from the process and calculate the sample p value and standard deviation. For illustration, assume that two defectives are observed in the sample. Thus:

$$p = \frac{f}{n} = \frac{2}{100} = [.02]$$

$$s_p = \sqrt{\frac{pq}{n}} = \sqrt{\frac{(.02)\,(.98)}{100}} = [.014]$$

Step 2: Determine the value of z for a two-sided test at a .05 level of significance. This value is 1.96.

Step 3: Determine the value of p_d (i.e., the displacement from p to be detected). For this example, the displacement is given as 1 standard deviation. Therefore, for a sample standard deviation of .014 and a sample p value of .02, the value of p_d is .02 plus .014, or .034.

Step 4: Given the stated values ($p = .02$, $p_d = .034$, $s_p = .014$, and $z = 1.96$), compute the required sample quantity.

$$n = \frac{(s_p)\,(z)^2}{(p_d - p)^2} = \frac{(.014)\,(1.96)^2}{(.034 - .02)^2} = [274]$$

Given the same conditions for a one-sided test, the required sample size is:

$$n = \frac{(s_p)\,(z)^2}{(p_d - p)^2} = \frac{(.014)\,(1.64)^2}{(.034 - .02)^2} = [192]$$

Exercise Worksheets

Exercise 8.1

TEST OF SIGNIFICANCE FOR MEANS (INDEPENDENT SAMPLES)

PROBLEM: Eight samples are drawn from an experimental process (process 1) and 6 samples are drawn from the presently used process (process 2). To a .05 level, determine if the mean of process 1 is significantly higher than process 2.

STEP 1: Given the following sample measurements, calculate the listed values for each process.

Process 1		Process 2	
16	19	20	2
9	10	5	4
4	5	1	
23	2	16	

$n_1 =$ [ ] $n_2 =$ [ ]

$\Sigma x_1 =$ [ ] $\Sigma x_2 =$ [ ]

$(\Sigma x_1)^2 =$ [ ] $(\Sigma x_2)^2 =$ [ ]

$\Sigma x_1^2 =$ [ ] $\Sigma x_2^2 =$ [ ]

$\bar{x}_1 =$ [ ] $\bar{x}_2 =$ [ ]

Exercise 8.1 (continued)

STEP 2: Using the following formula, calculate the combined sample variance.

$$s^2_{1,2} = \frac{\left[\Sigma x_1^{\,2} - \frac{(\Sigma x_1)^2}{n_1}\right] + \left[\Sigma x_2^{\,2} - \frac{(\Sigma x_2)^2}{n_2}\right]}{n_1 + n_2 - 2}$$

$$= \frac{\left[[\quad] - \frac{[\quad]}{[\quad]}\right] + \left[[\quad] - \frac{[\quad]}{[\quad]}\right]}{[\quad]}$$

$$= [\quad]$$

STEP 3: Calculate the t value and determine if significant.

$$t = \frac{\bar{x}_1 - \bar{x}_2}{\sqrt{\frac{s^2_{1,2}}{n_1} + \frac{s^2_{1,2}}{n_2}}}$$

$$= \frac{[\quad] - [\quad]}{\sqrt{\frac{[\quad]}{[\quad]} + \frac{[\quad]}{[\quad]}}} = [\quad]$$

df = [ ] critical t = [ ]

Exercise 8.2

TEST OF SIGNIFICANCE FOR MEANS
(CORRELATED SAMPLES)

PROBLEM: Ten samples are tested before and after burn-in. To a .05 level of significance, determine if a difference exists between the before-after results.

SAMPLE	BEFORE (x_1)	AFTER (x_2)	D ($x_1 - x_2$)	D^2
1	76	75	1	1
2	79	83	−4	16
3	72	75	−3	9
4	74	78	−4	16
5	73	71	2	4
6	72	75	−3	9
7	74	76	−2	4
8	75	74	1	1
9	78	81	−3	9
10	73	75	−2	4
			$\Sigma D = -17$	$\Sigma D^2 = 73$

Exercise 8.2 (continued)

$$t = \frac{\Sigma D}{\sqrt{\dfrac{(n)(\Sigma D^2) - (\Sigma D)^2}{n - 1}}}$$

$$t = \frac{[\quad\quad]}{\sqrt{\dfrac{[\quad\quad] \times [\quad\quad] - [\quad\quad]}{[\quad\quad]}}}$$

$$t = [\quad\quad]$$

Degrees of Freedom = [ ]

Critical Value of t = [ ]

Significant ☐ Yes ☐ No

Exercise 8.3

TEST OF SIGNIFICANCE FOR MEANS
(UNEQUAL POPULATION VARIANCE)

PROBLEM: Machines 1 and 2 perform similar operations. To a .05 level of significance, determine if a difference exists between the means of the two machines.

Machine 1	Machine 2
$n_1 = 19$	$n_2 = 13$
$\bar{x}_1 = 8.21$	$\bar{x}_2 = 6.15$
$\Sigma(x - \bar{x}_1)^2 = 87.08$	$\Sigma(x - \bar{x}_2)^2 = 17.66$

$$s_{\bar{x}_1}^{\ 2} = \frac{\Sigma(x - \bar{x}_1)^2}{n_1(n_1 - 1)} = \frac{[\qquad]}{[\qquad] \times [\qquad]} = [\qquad\qquad]$$

$$s_{\bar{x}_2}^{\ 2} = \frac{\Sigma(x - \bar{x}_2)^2}{n_2(n_2 - 1)} = \frac{[\qquad]}{[\qquad] \times [\qquad]} = [\qquad\qquad]$$

Exercise 8.3 (continued)

$$s_{\bar{x}_1-\bar{x}_2} = \sqrt{s_{\bar{x}_1}^{2} + s_{\bar{x}_2}^{2}} = \sqrt{[\quad] + [\quad]}$$

$$= [\quad]$$

$$t = \frac{\bar{x}_1 - \bar{x}_2}{s_{\bar{x}_1-\bar{x}_2}} = \frac{[\quad] - [\quad]}{[\quad]} = [\quad]$$

Critical Value of t for n_1 and n_2 (t_1 and t_2)

n_1 = [ ] df = [ ] t_1 = [ ]

n_2 = [ ] df = [ ] t_2 = [ ]

Critical Adjusted Value of t: (t_a)

$$t_a = \sqrt{\frac{(s_{\bar{x}_1}^{2})(t_1) + (s_{\bar{x}_2}^{2})(t_2)}{s_{\bar{x}_1}^{2} + s_{\bar{x}_2}^{2}}}$$

$$= \sqrt{\frac{[\quad] \times [\quad] + [\quad] \times [\quad]}{[\quad] + [\quad]}}$$

= [ ] Significant: ☐ Yes ☐ No

Exercise 8.4

TEST OF SIGNIFICANCE FOR PROPORTIONS
(INDEPENDENT SAMPLES)

PROBLEM: Operator A and B perform similar tasks. To a .05 level of significance, determine if a difference exists in the number of defects produced by each operator.

Operator A	Operator B
$n_1 = 170$	$n_2 = 150$
$f_1 = 5$	$f_2 = 8$
$p_1 = .029$	$p_2 = .053$

Exercise 8.4 (continued)

$$p_{1,2} = \frac{f_1 + f_2}{n_1 + n_2} = \frac{[\qquad] + [\qquad]}{[\qquad] + [\qquad]}$$

$$= [\qquad]$$

$$q = 1 - p_{1,2} = [\qquad]$$

$$s_{p_1 - p_2} = \sqrt{(p_{1,2})(q) \left(\frac{1}{n_1} + \frac{1}{n_2}\right)}$$

$$= \sqrt{([\qquad] \times [\qquad])([\qquad] + [\qquad])}$$

$$= [\qquad]$$

$$Z = \frac{p_1 - p_2}{s_{p_1 - p_2}} = \frac{[\qquad] - [\qquad]}{[\qquad]} = [\qquad]$$

Critical z Value = [ ]

Significant ☐ Yes ☐ No

Exercise 8.5

TEST OF SIGNIFICANCE FOR PROPORTIONS
(CORRELATED SAMPLES)

PROBLEM: One hundred products are tested on machine 1 and then tested on machine 2. To a .05 level of significance, determine if a difference exists between the two machines.

CONDITION	QUANTITY	PROPORTION
Pass Mach. 1/Fail Mach. 2	13	[ ]
Pass Mach. 1/Pass Mach. 2	53	[ ]
Fail Mach. 1/Fail Mach. 2	26	[ ]
Fail Mach. 1/Pass Mach. 2	8	[ ]

Exercise 8.5 (continued)

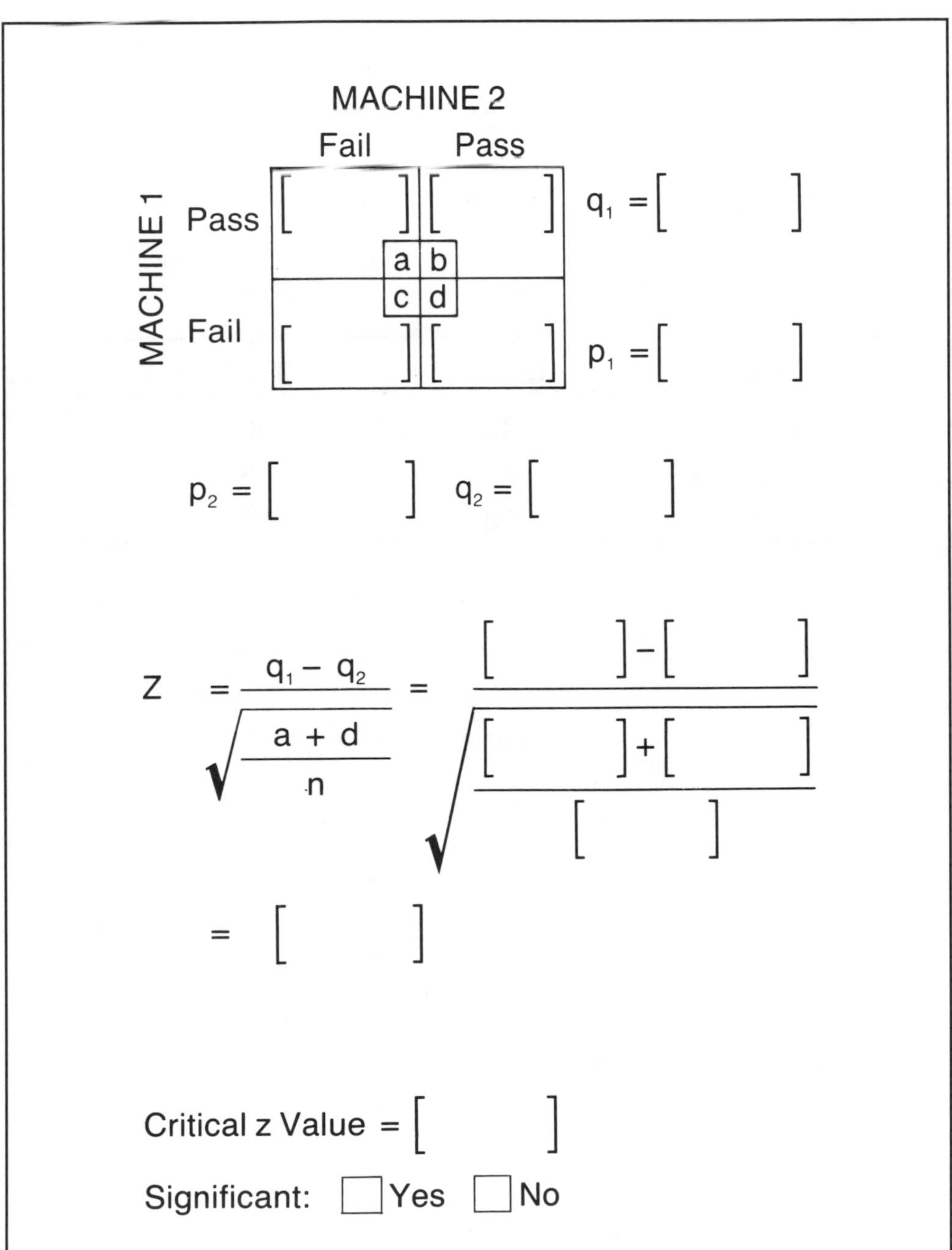

Exercise 8.6

TEST OF SIGNIFICANCE FOR FREQUENCIES
(INDEPENDENT SAMPLES)

PROBLEM: Site A and B are producing the same products. To a .05 level of significance, determine if site B has a higher outgoing quality level.

CATEGORY	SITE A	SITE B
Inspected	90	80
Accepted	60	65
Rejected	30	15

Exercise 8.6 (continued)

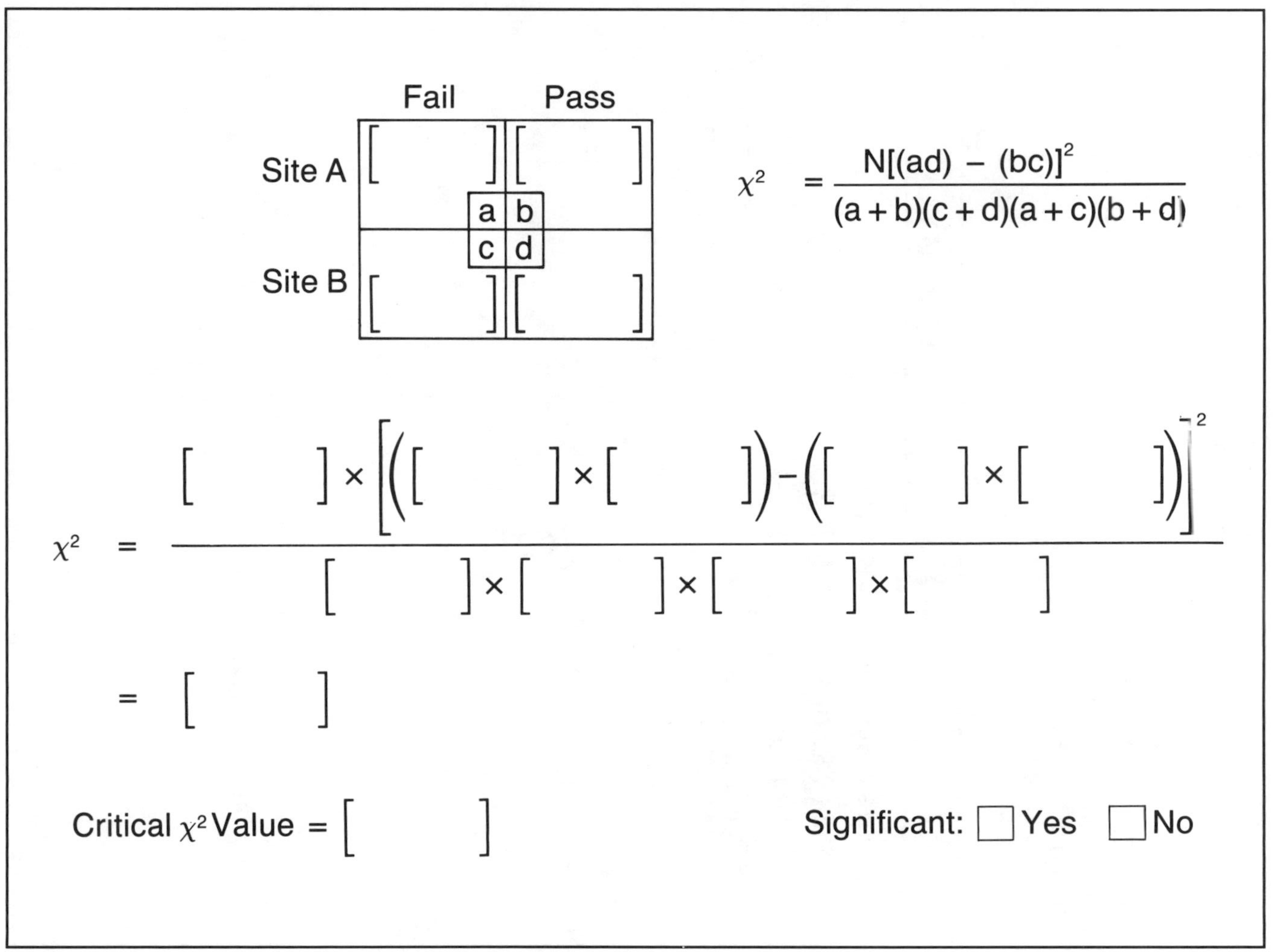

Exercise 8.7

TEST OF SIGNIFICANCE FOR FREQUENCIES
(CORRELATED SAMPLES)

PROBLEM: One hundred products are tested on machine 1 and then tested on machine 2. To a .05 level of significance, determine if a difference exists between the two machines.

CONDITION	QUANTITY
Pass Mach. 1/Fail Mach. 2	13
Pass Mach. 1/Pass Mach. 2	53
Fail Mach. 1/Fail Mach. 2	26
Fail Mach. 1/Pass Mach. 2	8

Exercise 8.7 (continued)

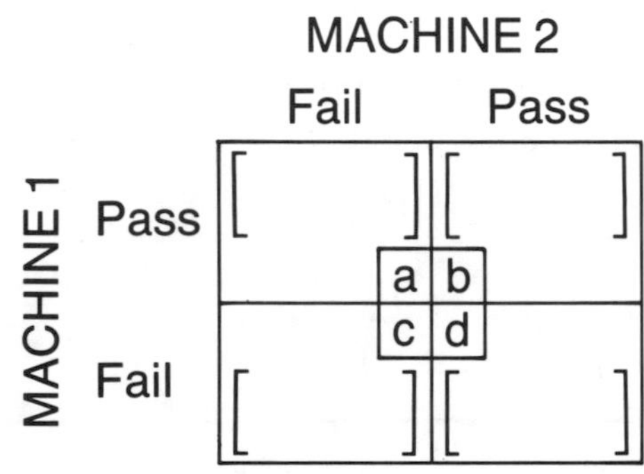

$$\chi^2 = \frac{(d - a)^2}{a + d} = \frac{([\quad] - [\quad])^2}{[\quad] + [\quad]}$$

$$= [\quad]$$

Critical χ^2 Value = [ ]

Significant: ☐ Yes ☐ No

Exercise 8.8

TEST OF SIGNIFICANCE FOR FREQUENCIES
(SAME-SAMPLE FREQUENCIES)

PROBLEM: One hundred products are tested on machine 1 and then tested on machine 2. To a level .05 of significance, determine if a difference exists between the two machines.

MACHINE NUMBER	QUANTITY		
	TESTED	PASS	FAIL
1	100	66	34
2	100	61	39

Exercise 8.8 (continued)

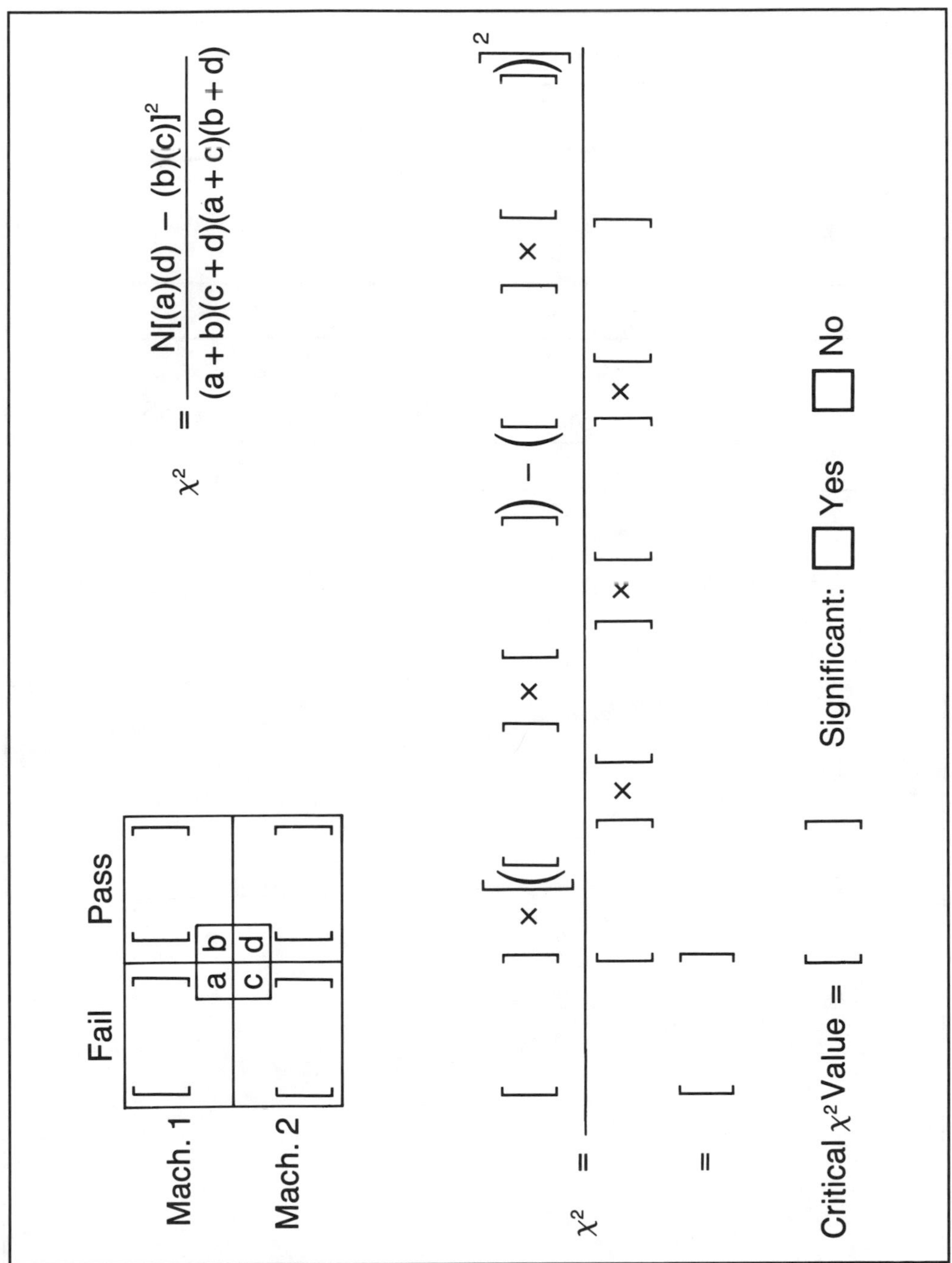

Exercise 8.9

TEST OF SIGNIFICANCE FOR VARIANCE
(INDEPENDENT SAMPLES)

PROBLEM: Supplier A and B produce the same products. To a .05 level of significance, determine if a difference in variation exists between the two suppliers.

Supplier A	Supplier B
$n_1 = 15$	$n_2 = 9$
$\Sigma(x - \bar{x}_1)^2 = 316$	$\Sigma(x - \bar{x}_2)^2 = 530$

Exercise 8.9 (continued)

$$s_1^2 = \frac{\Sigma(x - \bar{x}_1)^2}{n_1 - 1} = \frac{[\quad\quad]}{[\quad\quad]} = [\quad\quad]$$

$$s_2^2 = \frac{\Sigma(x - \bar{x}_2)^2}{n_2 - 1} = \frac{[\quad\quad]}{[\quad\quad]} = [\quad\quad]$$

$$F = \frac{s_x^2}{s_x^2} = \frac{[\quad\quad]}{[\quad\quad]} = [\quad\quad]$$

df for greater variance estimate = [ ]

df for lesser variance estimate = [ ]

Critical F Value = [ ]

Significant: ☐ Yes ☐ No

Exercise 8.10

TEST OF SIGNIFICANCE FOR VARIANCE
(CORRELATED SAMPLES)

PROBLEM: Twenty-five samples are tested before and after burn-in. To a .05 level of significance, determine if a difference in variance exists between the before-after test results.

Before	After
$n_1 = 25$	$n_2 = 25$
$\Sigma(x - \bar{x}_1)^2 = 212$	$\Sigma(x - \bar{x}_2)^2 = 364$

Correlation (r) = 0.85

Exercise 8.10 (continued)

$$s_1^2 = \frac{\Sigma(x - \bar{x}_1)^2}{n_1 - 1} = \frac{[\quad]}{[\quad]} = [\quad]$$

$$s_2^2 = \frac{\Sigma(x - \bar{x}_2)^2}{n_2 - 1} = \frac{[\quad]}{[\quad]} = [\quad]$$

$$t = \frac{\left(s_1^2 - s_2^2\right)\left(\sqrt{n - 2}\right)}{\sqrt{(4)(s_1^2)(s_2^2)(1 - r^2)}}$$

$$= \frac{\left([\quad] - [\quad]\right)\left([\quad]\right)}{\sqrt{[\quad] \times [\quad] \times [\quad] \times [\quad]}}$$

$$= [\quad]$$

df = [ ] Critical t Value = [ ]

Significant ☐ Yes ☐ No

Exercise 8.11

SAMPLE SIZE DETERMINATION
(FOR MEANS)

PROBLEM: For a one-sided test at a .05 level of significance, determine the minimum sample size to detect a 1 standard deviation displacement.

Sample Mean = 39.7
Standard Deviation = 1.55

Exercise 8.11 (continued)

$$x_d = \bar{x} + (d)(s)$$

$$= [\quad\quad] + ([\quad\quad] \times [\quad\quad])$$

$$= [\quad\quad]$$

$$n = \frac{(s)^2 \times (2z)^2}{(x_d - \bar{x})^2}$$

$$= \frac{[\quad\quad]^2 \times [[\quad\quad] \times [\quad\quad]]^2}{[[\quad\quad] - [\quad\quad]]^2}$$

$$= \frac{[\quad\quad] \times [\quad\quad]}{[\quad\quad]} = [\quad\quad]$$

Exercise 8.12

SAMPLE SIZE DETERMINATION
(FOR PROPORTIONS)

PROBLEM: For a two-sided test at a .05 level of significance (based on the observation of 4 defectives in a sample of 100), determine the minimum sample size to detect a displacement of 2 standard deviations.

Exercise 8.12 (continued)

$$p = \frac{f}{n} = \frac{[\quad\quad]}{[\quad\quad]} = [\quad\quad]$$

$$q = 1 - p = [\quad\quad]$$

$$s_p = \sqrt{\frac{pq}{n}} = \sqrt{\frac{[\quad\quad] \times [\quad\quad]}{[\quad\quad]}}$$

$$= [\quad\quad]$$

$$p_d = p + (d)(s_p)$$

$$= [\quad\quad] + \left([\quad\quad] \times [\quad\quad]\right) = [\quad\quad]$$

$$n = \frac{s_p\,(z)^2}{(p_d - p)^2} = \frac{[\quad\quad] \times [\quad\quad]^2}{[\quad\quad]^2}$$

$$= [\quad\quad]$$

TABLE 8.2 Critical Values of t

	Level of significance for one-tailed test					
	.10	.05	.025	.01	.005	.0005
	Level of significance for two-tailed test					
df	.20	.10	.05	.02	.01	.001
1	3.078	6.314	12.706	31.821	63.657	636.619
2	1.886	2.920	4.303	6.965	9.925	31.598
3	1.638	2.353	3.182	4.541	5.841	12.941
4	1.533	2.132	2.776	3.747	4.604	8.610
5	1.476	2.015	2.571	3.365	4.032	6.859
6	1.440	1.943	2.447	3.143	3.707	5.959
7	1.415	1.895	2.365	2.998	3.499	5.405
8	1.397	1.860	2.306	2.896	3.355	5.041
9	1.383	1.833	2.262	2.821	3.250	4.781
10	1.372	1.812	2.228	2.764	3.169	4.587
11	1.363	1.796	2.201	2.718	3.106	4.437
12	1.356	1.782	2.179	2.681	3.055	4.318
13	1.350	1.771	2.160	2.650	3.012	4.221
14	1.345	1.761	2.145	2.624	2.977	4.140
15	1.341	1.753	2.131	2.602	2.947	4.073
16	1.337	1.746	2.120	2.583	2.921	4.015
17	1.333	1.740	2.110	2.567	2.898	3.965
18	1.330	1.734	2.101	2.552	2.878	3.922
19	1.328	1.729	2.093	2.539	2.861	3.883
20	1.325	1.725	2.086	2.528	2.845	3.850
21	1.323	1.721	2.080	2.518	2.831	3.819
22	1.321	1.717	2.074	2.508	2.819	3.792
23	1.319	1.714	2.069	2.500	2.807	3.767
24	1.318	1.711	2.064	2.492	2.797	3.745
25	1.316	1.708	2.060	2.485	2.787	3.725
26	1.315	1.706	2.056	2.479	2.779	3.707
27	1.314	1.703	2.052	2.473	2.771	3.690
28	1.313	1.701	2.048	2.467	2.763	3.674
29	1.311	1.699	2.045	2.462	2.756	3.659
30	1.310	1.697	2.042	2.457	2.750	3.646
40	1.303	1.684	2.021	2.423	2.704	3.551
60	1.296	1.671	2.000	2.390	2.660	3.460
120	1.289	1.658	1.980	2.358	2.617	3.373
∞	1.282	1.645	1.960	2.326	2.576	3.291

TABLE 8.3 Critical Values of Chi Square (χ^2)

LEVEL OF SIGNIFICANCE

df	.99	.98	.95	.90	.80	.70	.50	.30	.20	.10	.05	.02	.01	.001
1	.00016	.00063	.0039	.016	.064	.15	.46	1.07	1.64	2.71	3.84	5.41	6.64	10.83
2	.02	.04	.10	.21	.45	.71	1.39	2.41	3.22	4.60	5.99	7.82	9.21	13.82
3	.12	.18	.35	.58	1.00	1.42	2.37	3.66	4.64	6.25	7.82	9.84	11.34	16.27
4	.30	.43	.71	1.06	1.65	2.20	3.36	4.88	5.99	7.78	9.49	11.67	13.28	18.46
5	.55	.75	1.14	1.61	2.34	3.00	4.35	6.06	7.29	9.24	11.07	13.39	15.09	20.52
6	.87	1.13	1.64	2.20	3.07	3.83	5.35	7.23	8.56	10.64	12.59	15.03	16.81	22.46
7	1.24	1.56	2.17	2.83	3.82	4.67	6.35	8.38	9.80	12.02	14.07	16.62	18.48	24.32
8	1.65	2.03	2.73	3.49	4.59	5.53	7.34	9.52	11.03	13.36	15.51	18.17	20.09	26.12
9	2.09	2.53	3.32	4.17	5.38	6.39	8.34	10.66	12.24	14.68	16.92	19.68	21.67	27.88
10	2.56	3.06	3.94	4.86	6.18	7.27	9.34	11.78	13.44	15.99	18.31	21.16	23.21	29.59
11	3.05	3.61	4.58	5.58	6.99	8.15	10.34	12.90	14.63	17.28	19.68	22.62	24.72	31.26
12	3.57	4.18	5.23	6.30	7.81	9.03	11.34	14.01	15.81	18.55	21.03	24.05	26.22	32.91
13	4.11	4.76	5.89	7.04	8.63	9.93	12.34	15.12	16.98	19.81	22.36	25.47	27.69	34.53
14	4.66	5.37	6.57	7.79	9.47	10.82	13.34	16.22	18.15	21.06	23.68	26.87	29.14	36.12
15	5.23	5.98	7.26	8.55	10.31	11.72	14.34	17.32	19.31	22.31	25.00	28.26	30.58	37.70
16	5.81	6.61	7.96	9.31	11.15	12.62	15.34	18.42	20.46	23.54	26.30	29.63	32.00	39.29
17	6.41	7.26	8.67	10.08	12.00	13.53	16.34	19.51	21.62	24.77	27.59	31.00	33.41	40.75
18	7.02	7.91	9.39	10.86	12.86	14.44	17.34	20.60	22.76	25.99	28.87	32.35	34.80	42.31
19	7.63	8.57	10.12	11.65	13.72	15.35	18.34	21.69	23.90	27.20	30.14	33.69	36.19	43.82
20	8.26	9.24	10.85	12.44	14.58	16.27	19.34	22.78	25.04	28.41	31.41	35.02	37.57	45.32
21	8.90	9.92	11.59	13.24	15.44	17.18	20.34	23.86	26.17	29.62	32.67	36.34	38.93	46.80
22	9.54	10.60	12.34	14.04	16.31	18.10	21.34	24.94	27.30	30.81	33.92	37.66	40.29	48.27
23	10.20	11.29	13.09	14.85	17.19	19.02	22.34	26.02	28.43	32.01	35.17	38.97	41.64	49.73
24	10.86	11.99	13.85	15.66	18.06	19.94	23.34	27.10	29.55	33.20	36.42	40.27	42.98	51.18
25	11.52	12.70	14.61	16.47	18.94	20.87	24.34	28.17	30.68	34.38	37.65	41.57	44.31	52.62
26	12.20	13.41	15.38	17.29	19.82	21.79	25.34	29.25	31.80	35.56	38.88	42.86	45.64	54.05
27	12.88	14.12	16.15	18.11	20.70	22.72	26.34	30.32	32.91	36.74	40.11	44.14	46.96	55.48
28	13.56	14.85	16.93	18.94	21.59	23.65	27.34	31.39	34.03	37.92	41.34	45.42	48.28	56.89
29	14.26	15.57	17.71	19.77	22.48	24.58	28.34	32.46	35.14	39.09	42.56	46.69	49.59	58.30
30	14.95	16.31	18.49	20.60	23.36	25.51	29.34	33.53	36.25	40.26	43.77	47.96	50.89	59.70

TABLE 8.4 Critical Values of F: 5% Level

df GREATER VARIANCE ESTIMATE (columns); df LESSER VARIANCE ESTIMATE (rows)

df LESSER VARIANCE ESTIMATE	1	2	3	4	5	6	7	8	9	10	12	15	20	24	30	40	60	120	∞
1	161.4	199.5	215.7	224.6	230.2	234.0	236.8	238.9	240.5	241.9	243.9	245.9	248.0	249.1	250.1	251.1	252.2	253.3	254.3
2	18.51	19.00	19.16	19.25	19.30	19.33	19.35	19.37	19.38	19.40	19.41	19.43	19.45	19.45	19.46	19.47	19.48	19.49	19.50
3	10.13	9.55	9.28	9.12	9.01	8.94	8.89	8.85	8.81	8.79	8.74	8.70	8.66	8.64	8.62	8.59	8.57	8.55	8.53
4	7.71	6.94	6.59	6.39	6.26	6.16	6.09	6.04	6.00	5.96	5.91	5.86	5.80	5.77	5.75	5.72	5.69	5.66	5.63
5	6.61	5.79	5.41	5.19	5.05	4.95	4.88	4.82	4.77	4.74	4.68	4.62	4.56	4.53	4.50	4.46	4.43	4.40	4.36
6	5.99	5.14	4.76	4.53	4.39	4.28	4.21	4.15	4.10	4.06	4.00	3.94	3.87	3.84	3.81	3.77	3.74	3.70	3.67
7	5.59	4.74	4.35	4.12	3.97	3.87	3.79	3.73	3.68	3.64	3.57	3.51	3.44	3.41	3.38	3.34	3.30	3.27	3.23
8	5.32	4.46	4.07	3.84	3.69	3.58	3.50	3.44	3.39	3.35	3.28	3.22	3.15	3.12	3.08	3.04	3.01	2.97	2.93
9	5.12	4.26	3.86	3.63	3.48	3.37	3.29	3.23	3.18	3.14	3.07	3.01	2.94	2.90	2.86	2.83	2.79	2.75	2.71
10	4.96	4.10	3.71	3.48	3.33	3.22	3.14	3.07	3.02	2.98	2.91	2.85	2.77	2.74	2.70	2.66	2.62	2.58	2.54
11	4.84	3.98	3.59	3.36	3.20	3.09	3.01	2.95	2.90	2.85	2.79	2.72	2.65	2.61	2.57	2.53	2.49	2.45	2.40
12	4.75	3.89	3.49	3.26	3.11	3.00	2.91	2.85	2.80	2.75	2.69	2.62	2.54	2.51	2.47	2.43	2.38	2.34	2.30
13	4.67	3.81	3.41	3.18	3.03	2.92	2.83	2.77	2.71	2.67	2.60	2.53	2.46	2.42	2.38	2.34	2.30	2.25	2.21
14	4.60	3.74	3.34	3.11	2.96	2.85	2.76	2.70	2.65	2.60	2.53	2.46	2.39	2.35	2.31	2.27	2.22	2.18	2.13
15	4.54	3.68	3.29	3.06	2.90	2.79	2.71	2.64	2.59	2.54	2.48	2.40	2.33	2.29	2.25	2.20	2.16	2.11	2.07
16	4.49	3.63	3.24	3.01	2.85	2.74	2.66	2.59	2.54	2.49	2.42	2.35	2.28	2.24	2.19	2.15	2.11	2.06	2.01
17	4.45	3.59	3.20	2.96	2.81	2.70	2.61	2.55	2.49	2.45	2.38	2.31	2.23	2.19	2.15	2.10	2.06	2.01	1.96
18	4.41	3.55	3.16	2.93	2.77	2.66	2.58	2.51	2.46	2.41	2.34	2.27	2.19	2.15	2.11	2.06	2.02	1.97	1.92
19	4.38	3.52	3.13	2.90	2.74	2.63	2.54	2.48	2.42	2.38	2.31	2.23	2.16	2.11	2.07	2.03	1.98	1.93	1.88
20	4.35	3.49	3.10	2.87	2.71	2.60	2.51	2.45	2.39	2.35	2.28	2.20	2.12	2.08	2.04	1.99	1.95	1.90	1.84
21	4.32	3.47	3.07	2.84	2.68	2.57	2.49	2.42	2.37	2.32	2.25	2.18	2.10	2.05	2.01	1.96	1.92	1.87	1.81
22	4.30	3.44	3.05	2.82	2.66	2.55	2.46	2.40	2.34	2.30	2.23	2.15	2.07	2.03	1.98	1.94	1.89	1.84	1.78
23	4.28	3.42	3.03	2.80	2.64	2.53	2.44	2.37	2.32	2.27	2.20	2.13	2.05	2.01	1.96	1.91	1.86	1.81	1.76
24	4.26	3.40	3.01	2.78	2.62	2.51	2.42	2.36	2.30	2.25	2.18	2.11	2.03	1.98	1.94	1.89	1.84	1.79	1.73
25	4.24	3.39	2.99	2.76	2.60	2.49	2.40	2.34	2.28	2.24	2.16	2.09	2.01	1.96	1.92	1.87	1.82	1.77	1.71
26	4.23	3.37	2.98	2.74	2.59	2.47	2.39	2.32	2.27	2.22	2.15	2.07	1.99	1.95	1.90	1.85	1.80	1.75	1.69
27	4.21	3.35	2.96	2.73	2.57	2.46	2.37	2.31	2.25	2.20	2.13	2.06	1.97	1.93	1.88	1.84	1.79	1.73	1.67
28	4.20	3.34	2.95	2.71	2.56	2.45	2.36	2.29	2.24	2.19	2.12	2.04	1.96	1.91	1.87	1.82	1.77	1.71	1.65
29	4.18	3.33	2.93	2.70	2.55	2.43	2.35	2.28	2.22	2.18	2.10	2.03	1.94	1.90	1.85	1.81	1.75	1.70	1.64
30	4.17	3.32	2.92	2.69	2.53	2.42	2.33	2.27	2.21	2.16	2.09	2.01	1.93	1.89	1.84	1.79	1.74	1.68	1.62
40	4.08	3.23	2.84	2.61	2.45	2.34	2.25	2.18	2.12	2.08	2.00	1.92	1.84	1.79	1.74	1.69	1.64	1.58	1.51
60	4.00	3.15	2.76	2.53	2.37	2.25	2.17	2.10	2.04	1.99	1.92	1.84	1.75	1.70	1.65	1.59	1.53	1.47	1.39
120	3.92	3.07	2.68	2.45	2.29	2.17	2.09	2.02	1.96	1.91	1.83	1.75	1.66	1.61	1.55	1.50	1.43	1.35	1.25
∞	3.84	3.00	2.60	2.37	2.21	2.10	2.01	1.94	1.88	1.83	1.75	1.67	1.57	1.52	1.46	1.39	1.32	1.22	1.00

TABLE 8.5 Critical Values of F: 10% Level

df GREATER VARIANCE ESTIMATE (columns); df LESSER VARIANCE ESTIMATE (rows)

	1	2	3	4	5	6	7	8	9	10	12	15	20	24	30	40	60	120	∞
1	39.86	49.50	53.59	55.83	57.24	58.20	58.91	59.44	59.86	60.19	60.71	61.22	61.74	62.00	62.26	62.53	62.79	63.06	63.33
2	8.53	9.00	9.16	9.24	9.29	9.33	9.35	9.37	9.38	9.39	9.41	9.42	9.44	9.45	9.46	9.47	9.47	9.48	9.49
3	5.54	5.46	5.39	5.34	5.31	5.28	5.27	5.25	5.24	5.23	5.22	5.20	5.18	5.18	5.17	5.16	5.15	5.14	5.13
4	4.54	4.32	4.19	4.11	4.05	4.01	3.98	3.95	3.94	3.92	3.90	3.87	3.84	3.83	3.82	3.80	3.79	3.78	3.76
5	4.06	3.78	3.62	3.52	3.45	3.40	3.37	3.34	3.32	3.30	3.27	3.24	3.21	3.19	3.17	3.16	3.14	3.12	3.10
6	3.78	3.46	3.29	3.18	3.11	3.05	3.01	2.98	2.96	2.94	2.90	2.87	2.84	2.82	2.80	2.78	2.76	2.74	2.72
7	3.59	3.26	3.07	2.96	2.88	2.83	2.78	2.75	2.72	2.70	2.67	2.63	2.59	2.58	2.56	2.54	2.51	2.49	2.47
8	3.46	3.11	2.92	2.81	2.73	2.67	2.62	2.59	2.56	2.54	2.50	2.46	2.42	2.40	2.38	2.36	2.34	2.32	2.29
9	3.36	3.01	2.81	2.69	2.61	2.55	2.51	2.47	2.44	2.42	2.38	2.34	2.30	2.28	2.25	2.23	2.21	2.18	2.16
10	3.29	2.92	2.73	2.61	2.52	2.46	2.41	2.38	2.35	2.32	2.28	2.24	2.20	2.18	2.16	2.13	2.11	2.08	2.06
11	3.23	2.86	2.66	2.54	2.45	2.39	2.34	2.30	2.27	2.25	2.21	2.17	2.12	2.10	2.08	2.05	2.03	2.00	1.97
12	3.18	2.81	2.61	2.48	2.39	2.33	2.28	2.24	2.21	2.19	2.15	2.10	2.06	2.04	2.01	1.99	1.96	1.93	1.90
13	3.14	2.76	2.56	2.43	2.35	2.28	2.23	2.20	2.16	2.14	2.10	2.05	2.01	1.98	1.96	1.93	1.90	1.88	1.85
14	3.10	2.73	2.52	2.39	2.31	2.24	2.19	2.15	2.12	2.10	2.05	2.01	1.96	1.94	1.91	1.89	1.86	1.83	1.80
15	3.07	2.70	2.49	2.36	2.27	2.21	2.16	2.12	2.09	2.06	2.02	1.97	1.92	1.90	1.87	1.85	1.82	1.79	1.76
16	3.05	2.67	2.46	2.33	2.24	2.18	2.13	2.09	2.06	2.03	1.99	1.94	1.89	1.87	1.84	1.81	1.78	1.75	1.72
17	3.03	2.64	2.44	2.31	2.22	2.15	2.10	2.06	2.03	2.00	1.96	1.91	1.86	1.84	1.81	1.78	1.75	1.72	1.69
18	3.01	2.62	2.42	2.29	2.20	2.13	2.08	2.04	2.00	1.98	1.93	1.89	1.84	1.81	1.78	1.75	1.72	1.69	1.66
19	2.99	2.61	2.40	2.27	2.18	2.11	2.06	2.02	1.98	1.96	1.91	1.86	1.81	1.79	1.76	1.73	1.70	1.67	1.63
20	2.97	2.59	2.38	2.25	2.16	2.09	2.04	2.00	1.96	1.94	1.89	1.84	1.79	1.77	1.74	1.71	1.68	1.64	1.61
21	2.96	2.57	2.36	2.23	2.14	2.08	2.02	1.98	1.95	1.92	1.87	1.83	1.78	1.75	1.72	1.69	1.66	1.62	1.59
22	2.95	2.56	2.35	2.22	2.13	2.06	2.01	1.97	1.93	1.90	1.86	1.81	1.76	1.73	1.70	1.67	1.64	1.60	1.57
23	2.94	2.55	2.34	2.21	2.11	2.05	1.99	1.95	1.92	1.89	1.84	1.80	1.74	1.72	1.69	1.66	1.62	1.59	1.55
24	2.93	2.54	2.33	2.19	2.10	2.04	1.98	1.94	1.91	1.88	1.83	1.78	1.73	1.70	1.67	1.64	1.61	1.57	1.53
25	2.92	2.53	2.32	2.18	2.09	2.02	1.97	1.93	1.89	1.87	1.82	1.77	1.72	1.69	1.66	1.63	1.59	1.56	1.52
26	2.91	2.52	2.31	2.17	2.08	2.01	1.96	1.92	1.88	1.86	1.81	1.76	1.71	1.68	1.65	1.61	1.58	1.54	1.50
27	2.90	2.51	2.30	2.17	2.07	2.00	1.95	1.91	1.87	1.85	1.80	1.75	1.70	1.67	1.64	1.60	1.57	1.53	1.49
28	2.89	2.50	2.29	2.16	2.06	2.00	1.94	1.90	1.87	1.84	1.79	1.74	1.69	1.66	1.63	1.59	1.56	1.52	1.48
29	2.89	2.50	2.28	2.15	2.06	1.99	1.93	1.89	1.86	1.83	1.78	1.73	1.68	1.65	1.62	1.58	1.55	1.51	1.47
30	2.88	2.49	2.28	2.14	2.05	1.98	1.93	1.88	1.85	1.82	1.77	1.72	1.67	1.64	1.61	1.57	1.54	1.50	1.46
40	2.84	2.44	2.23	2.09	2.00	1.93	1.87	1.83	1.79	1.75	1.71	1.66	1.61	1.57	1.54	1.51	1.47	1.42	1.38
60	2.79	2.39	2.18	2.04	1.95	1.87	1.82	1.77	1.74	1.71	1.66	1.60	1.54	1.51	1.48	1.44	1.40	1.35	1.29
120	2.75	2.35	2.13	1.99	1.90	1.82	1.77	1.72	1.68	1.65	1.60	1.55	1.48	1.45	1.41	1.37	1.32	1.26	1.19
∞	2.71	2.30	2.08	1.94	1.85	1.77	1.72	1.67	1.63	1.60	1.55	1.49	1.42	1.38	1.34	1.30	1.24	1.17	1.00

SECTION 9

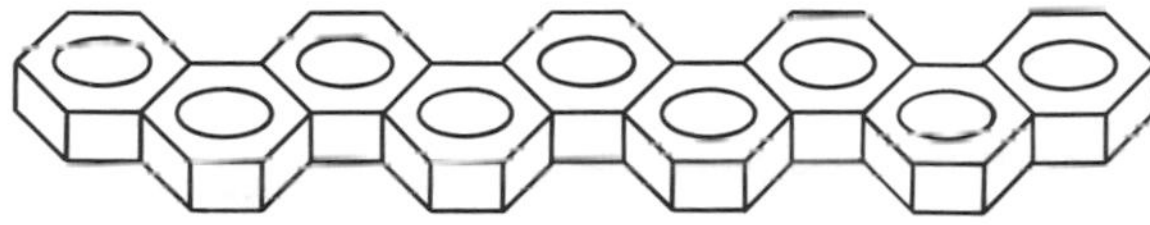

DESIGN OF EXPERIMENTS

EXPERIMENTAL DESIGNS

The design of an experiment is simply the process of determining how the experiment will be conducted. For example, the investigator must (1) select the values or categories of the independent variable, or variables, to be compared; (2) select samples for the experiment; (3) define rules or procedures whereby samples are assigned to particular independent variable categories; (4) specify the type of observations or measurements to be made on each sample and; (5) specify the statistical method(s) for comparing the experimental results.

The basic structure of many experiments involves a single independent variable at two or more levels. One such design is where the samples are divided into a control group and an experimental group. Comparisons are then made to determine if a significant difference resulted from the treatment condition. Another form of the single variable experiment is when both groups are exposed to different levels of the same treatment. While simpler in construction, single variable experiments are time-consuming and, therefore, costly. A more economical approach, while more complex, is factorial designs.

INDEPENDENT AND DEPENDENT VARIABLES

All experiments are concerned with the relationship between variables. The simplest type of experiment involves only two variables, an independent variable and a dependent variable.

An independent variable is a variable that is varied under controlled conditions by the experimenter. For the design of an experiment, an independent variable must be subjected to at least two levels of treatment. For example, to test the effect of thermal expansion, expansion must be measured under two different temperature conditions. In this case, temperature is the independent variable which is varied and controlled.

A dependent variable is a variable whose value or condition is directly or indirectly affected by the independent variable. For example, the degree of thermal expansion (the dependent variable) is dependent upon temperature (the independent variable). In this manner, the relationship between the independent and dependent variables is defined through experimentation.

Any independent variable is referred to as a factor. Experiments which investigate the effects of two independent variables simultaneously are called two-factor or two-way classification experiments. When three independent variables are involved, the experiment is called a three-factor experiment, etc.

The different values or categories of the independent variable are called levels. Therefore, there may be two, three, or more levels for each factor.

RANDOMIZATION

In experimental designs, the sequence in which items are subjected to various conditions are often randomized to control for undefined variables that might bias the results.

An effective method of randomizing items, or groups of items, is to assign numbers and then use a table of random numbers to select the events or sequence of events in which the samples will be subjected.

Another form of randomization, one used for factorial experiments, is a randomized block design. This design assures, to the extent possible, that sample groups are homogeneous prior to experimentation. For example, in an experiment to determine optimal terminal crimping pressure versus the angle of the crimping tool (for maximum terminal/lead pull-strength), suppose the first sample group, say for low pressure/wide angle, contained terminals from supplier A and wire from supplier 1; the second sample for low pressure/narrow angle contained terminals from supplier B and wire from supplier 2; the third sample for high pressure/wide angle contained terminals from supplier A but wire from supplier 2; and the fourth sample for high pressure/narrow angle contained terminals from supplier B and wire from supplier 1.

If an experiment was run in this fashion, it would be impossible to determine whether good or poor results were due to pressure, angle, terminal supplier A or B, wire supplier 1 or 2, or any combination thereof. In other words, there are more extraneous independent variables than assigned independent variables.

The randomized block design, therefore, is a design that blocks the extraneous variables through randomized sample selection. For example, in the above experiment samples for the four sample groups would be randomly selected from suppliers A, B, 1, and 2. Then, to complete the randomization process the sample groups would be randomly assigned to the four experimental categories (i.e., low pressure/wide angle, low pressure/narrow angle, etc.). To further block for possible extraneous variables, the sequence in which the experimental category is run would also be randomized.

SINGLE FACTOR EXPERIMENTS

A single factor experiment is an experiment where only one independent variable is subjected to two or more levels.

Single Group Design: In this design, a single group of samples are sequentially subjected to different conditions or levels. Measurements are taken after each condition, one which is usually the normal condition, and compared for significance. If individual sample measurements are paired, a high correlation will usually exist between any two before/after conditions.

Two Group Design: In this design, a single sample group is randomly divided into two independent subgroups. Each subgroup is then subject to separate conditions and the results compared. In the two group design, one subgroup may be a control where no experimental conditions are applied.

Matched Group Design: In this design, samples are matched on one variable and then, after treatment, measured on another variable. This design can be either a single group or two group design.

ANALYSIS OF VARIANCE: ONE-WAY CLASSIFICATION

Purpose

Analysis of variance is a method for dividing the variation observed in experimental data into different parts, with each part assignable to a known source, cause, or factor. The magnitude of the variation resulting from the different sources is then compared to the F distribution to determine whether a particular part of the variation is greater than expected, or significant.

Partitioning Sum of Squares

The total sum of squares, or total variation, for two or more distributions is comprised of both the within sample group variation and the between sample group variation. In keeping with analysis of variance terminology, this is called the within-group sum of squares (SS_w) and the between-group sum of squares (SS_b). The combination of these two sums of squares is the total sum of squares (SS_t).

This division of the sum of squares identifies the deviation of a particular score from the mean to which the score belongs $(x_1 - \bar{x})^2$ and also the deviation of the group mean from the grand mean $(\bar{x}_1 - \bar{\bar{x}})^2$.

DEGREES OF FREEDOM

Each of the three sum of squares has an associated number of degrees of freedom.

The degrees of freedom for the total sum of squares, since it is comprised of each sample in each sample group, is the total number of all samples minus 1 $(N - 1)$.

The degrees of freedom for the within-group sum of squares is, for each group, the number of samples in each group minus 1. Thus, the number of degrees of freedom is the total of all samples for all groups minus the number of groups $(N - k)$. Where k is the symbol for the number of groups.

The number of degrees of freedom associated with the between-groups sum of squares is simply the number of groups minus 1 $(k - 1)$.

Since the within-group sum of squares and between-group sum of squares equal the total sum of squares, the within-group and between-group degrees of freedom must also equal the total degrees of freedom. Thus:

$$N - 1 = (N - k) + (k - 1)$$

$$\text{Total} = \text{Within} + \text{Between}$$

VARIANCE ESTIMATES

The between-groups variance estimate is a measure of variance attributable to the experimental treatments, while the within-group variance estimate is a measure of variance attributable to random chance factors. Therefore, the F value, which is obtained by dividing the between-groups estimate by the within-group estimate, is a measure of the significance of the experimental treatment.

If there is no real treatment effect, the between-group estimate and within-group estimate will be approximately the same; an F value of 1. As more and more of the variation is attributable to the experimental treatments, the ratio between the between-groups and within-group variance estimates will increase (a larger between-groups variance), causing the F ratio to increase; thereby increasing the probability that the treatment effect was responsible for the difference in the observed variance.

The between-groups variance estimate and the within-group variance estimate are obtained by dividing their sum of squares values by their degrees of freedom. This is expressed algebraically as:

$$s_b^2 = \frac{ss_b}{df} \text{ and } s_w^2 = \frac{ss_w}{df}$$

where s_b^2 and s_w^2 are the between-groups and within-group variance estimates.

The F ratio, or value, is then obtained by dividing the between-groups' estimate by the within-group's estimate:

$$F = \frac{s_b^2}{s_w^2}$$

Computation Formulas

Between-Groups Sum of Squares: $ss_b = \Sigma \frac{(\Sigma x)^2}{n} - \frac{(\Sigma\Sigma x)^2}{N}$

Within-Groups Sum of Squares: $ss_w = \Sigma\Sigma x^2 - \Sigma \frac{(\Sigma x)^2}{n}$

Total Sum of Squares: $ss_t = \Sigma\Sigma x^2 - \frac{(\Sigma\Sigma x)^2}{N}$

Between-Groups Variance Estimate: $s_b^2 = \frac{ss_b}{df}$

Within-Groups Variance Estimate: $s_w^2 = \frac{ss_w}{df}$

F Ratio: $F = \frac{s_b^2}{s_w^2}$

Example 9.1

A printer experiment involving four impact pressure levels was conducted to determine if impact pressure, through the normal range of adjustment, would significantly affect print clarity (assume a .05 level of significance).

SAMPLE GROUP			
I	II	III	IV
x	x	x	x
5	9	8	1
7	11	6	3
6	8	9	4
3	7	5	5
9	7	7	1
7		4	4
4		4	
2			

Step 1: Calculate specified values for each group.

		GROUP			
		I	II	III	IV
n	=	8	5	7	6
Σx	=	43	42	43	18
$\frac{(\Sigma x)^2}{n}$	=	231.13	352.80	264.14	54.00
Σx^2	=	269	364	287	68

Step 2: Sum the above values across all groups and calculate $(\Sigma\Sigma x)^2/N$.

$$N = 26$$

$$\Sigma\Sigma x = 146 \qquad \frac{(\Sigma\Sigma x)^2}{N} = 819.84$$

$$\Sigma\frac{(\Sigma x)^2}{n} = 902.07$$

$$\Sigma\Sigma x^2 = 988$$

Step 3: Calculate the sum of squares.

$$ss_b = \Sigma \frac{(\Sigma x)^2}{n} - \frac{(\Sigma\Sigma x)^2}{N}$$

$$= 902.07 - 819.84 = [82.23]$$

$$ss_w = \Sigma\Sigma x^2 - \Sigma \frac{(\Sigma x)^2}{n}$$

$$= 988 - 902.07 = [85.93]$$

$$ss_t = \Sigma\Sigma x^2 - \frac{(\Sigma\Sigma x)^2}{N}$$

$$= 988 - 819.84 = [168.16]$$

Step 4: Determine degrees of freedom.

$$df_b = 4 \text{ groups} - 1 \text{ group} = 4 - 1 = [3]$$

$$df_w = [n_1 - 1] + [n_2 - 1] + [n_3 - 1] + [n_4 - 1]$$
$$= [8 - 1] + [5 - 1] + [7 - 1] + [6 - 1] = [22]$$

$$df_t = N - 1 = 26 - 1 = [25]$$

Step 5: Calculate variance estimates.

$$s_b{}^2 = \frac{ss_b}{df} = \frac{82.23}{3} = [27.41]$$

$$s_w{}^2 = \frac{ss_w}{df} = \frac{85.93}{22} = [3.90]$$

Step 6: Calculate F ratio.

$$F = \frac{s_b{}^2}{s_w{}^2} = \frac{27.41}{3.90} = [7.02]$$

Step 7: Determine critical F value.

Refer to F table (Table 8.4, page 204) for .05 level of significance with 3 df for greater variance estimate and 22 df for lesser variance estimate. The critical F value is 3.05. Therefore, impact pressure, through the normal range of adjustment, has a significant effect on print clarity.

SOURCE OF VARIATION	SUM OF SQUARES	DEGREES OF FREEDOM	VARIANCE ESTIMATE	F VALUE
BETWEEN	82.23	3	27.41	7.02
WITHIN	85.93	22	3.90	
TOTAL	168.16	25		

FACTORIAL EXPERIMENTS

Structure

As opposed to single factor experiments where the effect of a single independent variable is studied, experiments can also be designed to study the effects of two or more independent variables simultaneously. For example, the effect of both temperature and humidity on corrosion; or for multiple dependent variables, the effect of temperature and humidity on corrosion, thermal leakage, and performance characteristics.

Such experiments are called factorial experiments. They are experiments where two or more factors under two or more levels are studied in combination. In this manner, the effects of two or more independent variables can be measured in a single experiment.

The factorial design provides maximum information with minimum expenditure of resources. For example, the required number of runs for a 3 variable/2 level experiment would be:

SINGLE RUN	SINGLE RUN WITH INTERACTION			FACTORIAL		
(one-way)	(one-way)	(two-way)	(three-way)	A	B	C
A+	A+	A+B+	A+B+C+	+	+	+
A−	A−	A+B−	A+B+C−	+	+	−
B+	B+	A−B+	A+B−C+	+	−	+
B−	B−	A−B−	A+B−C−	+	−	−
C+	C+	A+C+	A−B+C+	−	+	+
C−	C−	A+C−	A−B+C−	−	+	−
		A−C+	A−B−C+	−	−	+
		A−C−	A−B−C−	−	−	−
		B+C+				
		B+C−				
		B−C+				
		B−C−				
6 runs		26 runs			8 runs	

In a two factorial design, as illustrated in Figure 9.1, each independent variable (A and B) has two levels (1 and 2). Thus, there are four possible combinations: $A_1 B_1$, $A_1 B_2$, $A_2 B_1$, and $A_2 B_2$.

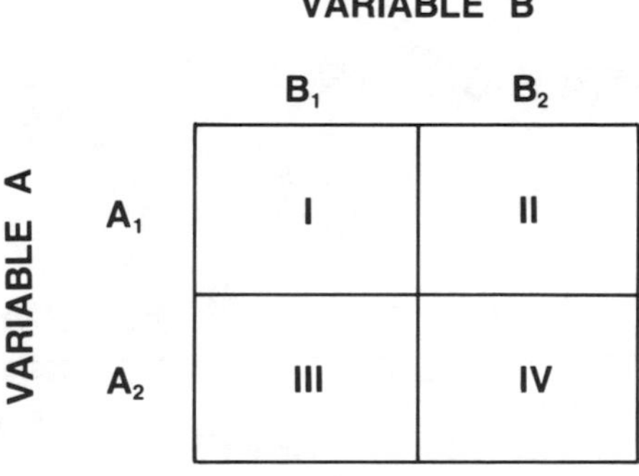

FIGURE 9.1

This design, since there are two levels for each factor (variables), is called a 2×2 factorial design. Factorial designs may also involve more levels or more factors (Figure 9.2).

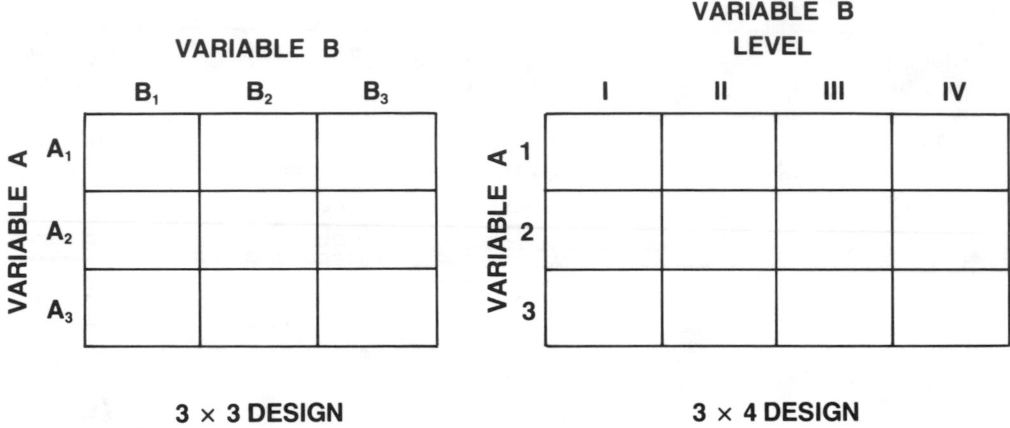

FIGURE 9.2

INTERACTION

The major advantage of a factorial design, other than economy, is that information is obtained about the interaction between variables (i.e., main effects, simple effects, and interaction effects).

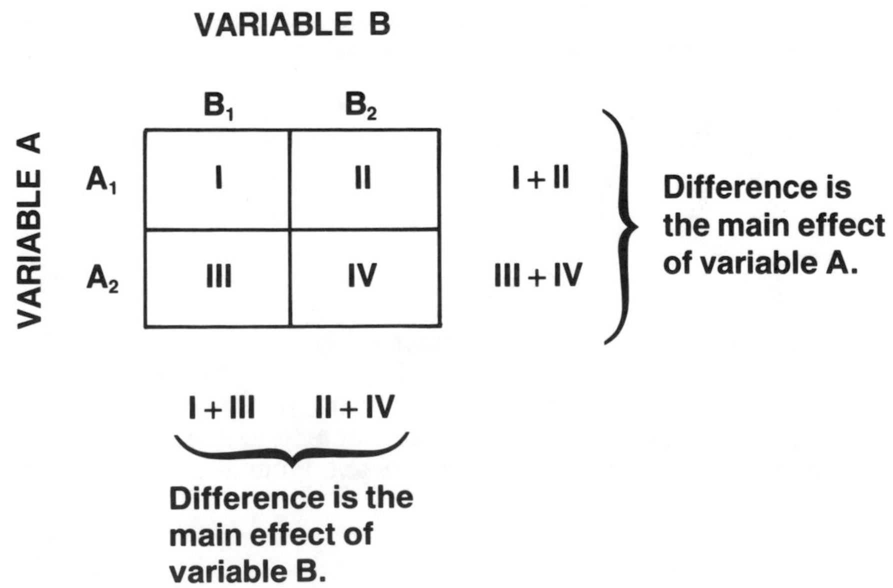

Simple Effects:	Effect of B under A_1	I-II
	Effect of B under A_2	III-IV
	Effect of A under B_1	I-III
	Effect of A under B_2	II-IV

Interaction Effects:	Interaction of B under A:	I-II - III-IV
	Interaction of A under B:	I-III - II-IV

To evaluate the effect of variable A, the average ($\bar{x}$) under condition A_1 is compared with the average under condition A_2 (i.e., the averages for quadrants I and II are summed and divided by 2 to obtain an average result under condition A_1. Likewise, the averages for quadrants III and IV are summed and divided by 2 to obtain an average result under condition A_2). The difference between the average result under condition A_1 and the average result under condition A_2 is the main effect of variable A.

The main effect of variable B is obtained in the same manner as variable A. The results in quadrants I and III are combined to determine the average result under condition B_1 and the results in quadrants II and IV are combined to determine the average results under condition B_2. The difference between B_1 and B_2 is the main effect of B.

The effects of the two levels of B under each level of A, and the effects of the two levels of A under each level of B are called the simple effects of the design. If variable A is temperature for example, and variable B is speed, the experimenter can determine the effect of different speeds at low and high temperatures, or the effect of different temperatures at low and high speeds.

The effect of both levels of A under both levels of B, or the effect of both levels of B under both levels of A, is called the interaction effect. **Note:** An interaction effect will only occur if the effect of one variable is dependent on the level of another variable. If the two variables are independent, an interaction effect will not occur.

The presence or absence of an interaction is determined by comparing the effect of variable A at the first level of variable B (I-II or the effect of B at A_1) with the effect of the B variable at the second level of variable A (III-IV or the effect of B at A_2). Accordingly, the effect of A at the first level of B (I-III) is compared with the effect of A at the second level of B (II-IV). Thus, the experimental results are examined to find any difference between differences. In other words, the simple effects are compared to determine if they are approximately equal or not. When they are not equal, an interaction exists between the two variables.

While the basic purpose of any experiment is to study the relationships between variable(s) under two or more conditions, the benefit of the factorial design is that all relationships can be studied simultaneously. Thereby eliminating unnecessary time and expense in conducting single experiments.

Example 9.2

In relation to trace peel strength for a particular type of printed circuit board, determine if an interaction exists between solder temperature (variable A) and belt speed (variable B). For this example, 4 samples each are tested under each condition.

Main Effects of:

$$A = \left| \bar{x}_{I} + \bar{x}_{II} \right| - \left| \bar{x}_{III} + \bar{x}_{IV} \right|$$

$$= \left| 81.5 + 70.5 \right| - \left| 48.5 + 57.5 \right| = 152 - 106 = 46$$

$$B = \left| \bar{x}_{I} + \bar{x}_{III} \right| - \left| \bar{x}_{II} + \bar{x}_{IV} \right|$$

$$= \left| 81.5 + 48.5 \right| - \left| 70.5 + 57.5 \right| = 130 - 128 = 2$$

Simple Effects of:

$$\text{B under } A_1 = \left| \bar{x}_{I} - \bar{x}_{II} \right| = \left| 81.5 - 70.5 \right| = 11$$

$$\text{B under } A_2 = \left| \bar{x}_{III} - \bar{x}_{IV} \right| = \left| 48.5 - 57.5 \right| = 9$$

$$\text{A under } B_1 = \left| \bar{x}_{I} - \bar{x}_{III} \right| = \left| 81.5 - 48.5 \right| = 33$$

$$\text{A under } B_2 = \left| \bar{x}_{II} - \bar{x}_{IV} \right| = \left| 70.5 - 57.5 \right| = 13$$

Interaction Effects of:

$$\text{B under A} = \left| \bar{x}_{I} - \bar{x}_{II} \right| - \left| \bar{x}_{III} - \bar{x}_{IV} \right|$$

$$= \left| 81.5 - 70.5 \right| - \left| 48.5 - 57.5 \right|$$

$$= 11 - 9 = 2$$

$$\text{A under B} = \left| \bar{x}_{I} - \bar{x}_{III} \right| - \left| \bar{x}_{II} - \bar{x}_{IV} \right|$$

$$= \left| 81.5 - 48.5 \right| - \left| 70.5 - 57.5 \right|$$

$$= 33 - 13 = 20$$

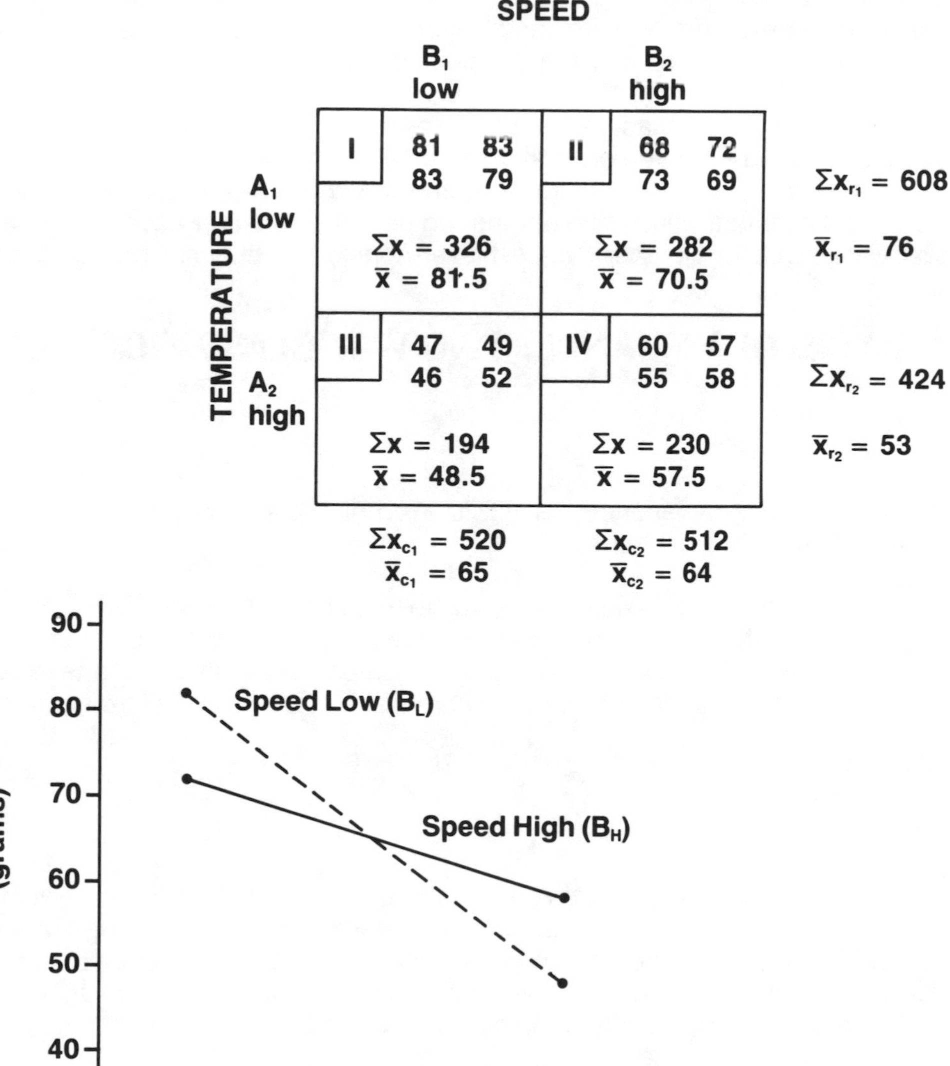

SPEED

TEMPERATURE	B_1 low	B_2 high	
A_1 low	I 81 83 83 79 $\Sigma x = 326$ $\bar{x} = 81.5$	II 68 72 73 69 $\Sigma x = 282$ $\bar{x} = 70.5$	$\Sigma x_{r_1} = 608$ $\bar{x}_{r_1} = 76$
A_2 high	III 47 49 46 52 $\Sigma x = 194$ $\bar{x} = 48.5$	IV 60 57 55 58 $\Sigma x = 230$ $\bar{x} = 57.5$	$\Sigma x_{r_2} = 424$ $\bar{x}_{r_2} = 53$
	$\Sigma x_{c_1} = 520$ $\bar{x}_{c_1} = 65$	$\Sigma x_{c_2} = 512$ $\bar{x}_{c_2} = 64$	

In this example, temperature has the main effect on peel strength. There is also an interaction between temperature and speed. Thus, all things considered, low temperature at low speed results in the best possible combination.

If a conventional single variable experiment was conducted to determine the effect of temperature or belt speed on peel strength, the procedure would be to set the temperature to the normal midpoint value and vary the belt speed, or set the belt speed to its midpoint value and vary the temperature. In either case, one would probably conclude that neither variable had an appreciable effect on peel strength.

ANALYSIS OF VARIANCE: TWO-WAY CLASSIFICATION

Purpose

A two-way analysis of variance permits the simultaneous investigation of two experimental variables.

To illustrate, assume that we wish to investigate the effects of two variables on print clarity. For example, hard and soft impact pressure under the condition of slow, medium, and fast print speeds. In this case, with impact pressure being one experimental variable and print speed the other, there are six combinations of experimental conditions.

Structure

The common method of conducting a two-way experiment is to randomly select a group of samples from the population and then to randomly assign an equal number of samples to each experimental condition. Therefore, for a 2×3 design, if a sample of 10 is selected, the total number of required samples would be $2 \times 3 \times 10$, or 60 (i.e., 10 samples for each of the six experimental conditions).

The resulting experimental data would then be arranged in a table containing two rows and three columns (Figure 9.3). The rows corresponding to the variable with the least experimental levels and the columns corresponding to the variable with the most experimental levels. For example, in the above print clarity illustration, impact pressure would correspond with the rows and print speed would correspond with the columns. Differences in the means of the rows would, therefore, result from differences in impact pressure and differences in the means of the columns resulting from differences in print speed.

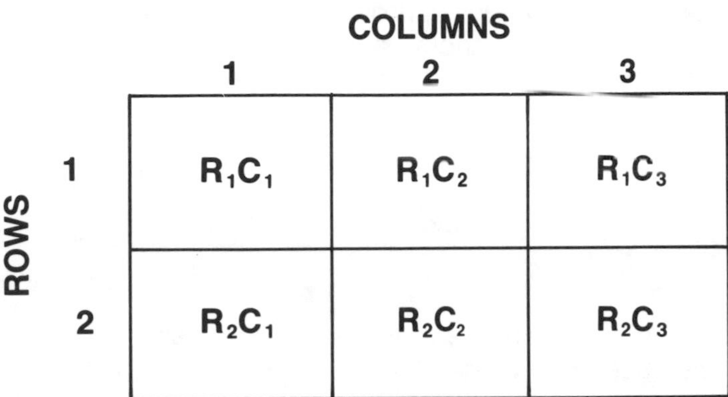

FIGURE 9.3

With this arrangement, the total sum of squares are partitioned into four parts: between-rows, between-columns, interaction, and within-cells. Then, since each sum of squares has an associated number of degrees of freedom, the four variance estimates are obtained by dividing each sum of squares' category by its associated degrees of freedom. The variance estimates are then used to test the significance of the difference between row means, column means, interaction effect, and within-cell variance.

Computation Formulas

Sum of Squares

Rows:

$$ss_r = \frac{(\Sigma x_{r_1})^2 + (\Sigma x_{r_2})^2}{n_r} - \frac{(\Sigma x_{r_1} + \Sigma x_{r_2})^2}{N}$$

where Σx_{r_1} is the sum of the X values in row 1, Σx_{r_2} is the sum of the X values in row 2, n_r is the number of samples, or values, in a given row, and N is the total of samples for both rows.

Columns:

$$ss_c = \frac{(\Sigma x_{c_1})^2 + (\Sigma x_{c_2})^2 + (\Sigma x_{c_3})^2}{n_c} - \frac{(\Sigma x_{c_1} + \Sigma x_{c_2} + \Sigma x_{c_3})^2}{N}$$

where Σx_{c_1} is the sum of the X values in column 1, Σx_{c_2} is the sum of the X values in column 2, Σx_{c_3} is the sum of the X values in column 3, n_c is the number of samples in a single column and N is the total number of samples in all columns.

Within Cells:

$$ss_w = \Sigma x_t^2 - \frac{\left[\begin{array}{l}(\Sigma x_{r_1c_1})^2 + (\Sigma x_{r_1c_2})^2 + (\Sigma x_{r_1c_3})^2 + \\ (\Sigma x_{r_2c_1})^2 + (\Sigma x_{r_2c_2})^2 + (\Sigma x_{r_2c_3})^2\end{array}\right]}{n_w}$$

where Σx_t^2 is the sum of the squared sample values for all samples, $\Sigma x_{r_1c_1}$ is the sum of the sample values in row 1 column 1, $\Sigma x_{r_1c_2}$ is the sum of the sample values in row 1 column 2, etc., and n_w is the number of samples in an individual cell.

Interaction:

$$ss_i = \frac{\left[\begin{array}{l}(\Sigma x_{r_1c_1})^2 + (\Sigma x_{r_1c_2})^2 + (\Sigma x_{r_1c_3})^2 + \\ (\Sigma x_{r_2c_1})^2 + (\Sigma x_{r_2c_2})^2 + (\Sigma x_{r_2c_3})^2\end{array}\right]}{n_w}$$

$$- \frac{\left[(\Sigma x_{r_1})^2 + (\Sigma x_{r_2})^2\right]}{n_r}$$

$$- \frac{\left[(\Sigma x_{c_1})^2 + (\Sigma x_{c_2})^2 + (\Sigma x_{c_3})^2\right]}{n_c}$$

$$+ \frac{(\Sigma x_t)^2}{N}$$

where $(\Sigma x_{r_1c_1})^2$ through $(\Sigma x_{r_2c_3})^2$ is the sum of the sample values squared for row 1 column 1 through row 2 column 3, where $(\Sigma x_{r_1})^2$ and $(\Sigma x_{r_2})^2$ are the sum of the sample values squared for row 1 and row 2, where $(\Sigma x_{c_1})^2$, $(\Sigma x_{c_2})^2$, and $(\Sigma x_{c_3})^2$ are the sum of the sample values squared for column 1, 2, and 3, where $(\Sigma x_t)^2$ is the sum of all sample values squared, where n_w is the number of samples in any one cell, n_r is the number of samples in any one row, n_c is the number of samples in any one column and N is the total of all samples.

Total:

$$ss_t = \Sigma x_t^2 - \frac{(\Sigma x_t)^2}{N}$$

where Σx_t^2 is the sum of the individual sample values squared, Σx_t is the sum of all sample values, and N is the total number of samples.

Degrees of Freedom

Rows: R − 1
Where R is the total number of rows.

Columns: C − 1
Where C is the total number of columns.

Interaction: (R −1)(C −1)
Where R is the total number of rows and C is the total number of columns.

Within Cells: N − (R)(C)
Where N is the total number of sample values, R is the total number of rows, and C is the total number of columns.

Total: N −1
Where N is the total number of samples.

Variance Estimate:

Rows

$$s_r^2 = \frac{ss_r}{df}$$

Columns

$$s_c^2 = \frac{ss_c}{df}$$

Within Cells

$$s_w^2 = \frac{ss_w}{df}$$

Interaction

$$s_i^2 = \frac{ss_i}{df}$$

where ss_r, ss_c, ss_w, and ss_i are the sum of squares for rows, columns, within cells, and interaction; and df is the associated degrees of freedom.

F Ratio

Rows

$$F_r = \frac{s_r^2}{s_w^2}$$

Columns

$$F_c = \frac{s_c^2}{s_w^2}$$

Interaction

$$F_i = \frac{s_i^2}{s_w^2}$$

Where s_r^2, s_c^2, s_i^2, and s_w^2 are the variance estimates.

Example 9.3

PURPOSE: To determine if print speed and/or impact pressure, or the interaction between the two, has a significant (.05 level) effect on print clarity.

DESIGN: The experimental design is to print 8 samples in each of 6 impact pressure/print speed combinations (i.e., a 2×3 design with 2 levels of impact pressure and 3 levels of print speed).

MEASUREMENT: Print clarity is measured in respect to a zero reference where smaller values represent better clarity.

Step 1: List data in matrix form and perform necessary calculations.

PRINT SPEED

IMPACT PRESSURE	Low	Medium	High	Total
Low	r_1c_1: 26, 14; 41, 16; 28, 29; 92, 31	r_1c_2: 41, 82; 26, 86; 19, 45; 59, 37	r_1c_3: 36, 87; 39, 99; 59, 126; 27, 104	
	$\Sigma x_{r_1c_1}$ = 277 $(\Sigma x_{r_1c_1})^2$ = 76,729 $\Sigma x^2_{r_1c_1}$ = 13,859	$\Sigma x_{r_1c_2}$ = 395 $(\Sigma x_{r_1c_2})^2$ = 156,025 $\Sigma x^2_{r_1c_2}$ = 23,713	$\Sigma x_{r_1c_3}$ = 577 $(\Sigma x_{r_1c_3})^2$ = 332,929 $\Sigma x^2_{r_1c_3}$ = 51,089	Σx_{r_1} = 1,249 $(\Sigma x_{r_1})^2$ = 1,560,001 $\Sigma x^2_{r_1}$ = 88,661
High	r_2c_1: 51, 35; 96, 36; 97, 28; 22, 76	r_2c_2: 39, 114; 104, 92; 130, 87; 122, 64	r_2c_3: 42, 133; 92, 124; 156, 68; 144, 142	
	$\Sigma x_{r_2c_1}$ = 441 $(\Sigma x_{r_2c_1})^2$ = 194,481 $\Sigma x^2_{r_2c_1}$ = 30,791	$\Sigma x_{r_2c_2}$ = 752 $(\Sigma x_{r_2c_2})^2$ = 565,504 $\Sigma x^2_{r_2c_2}$ = 77,246	$\Sigma x_{r_2c_3}$ = 901 $(\Sigma x_{r_2c_3})^2$ = 811,801 $\Sigma x^2_{r_2c_3}$ = 113,153	Σx_{r_2} = 2,094 $(\Sigma x_{r_2})^2$ = 4,384,836 $\Sigma x^2_{r_2}$ = 221,190
Total	Σx_{c_1} = 718 $(\Sigma x_{c_1})^2$ = 515,524 $\Sigma x^2_{c_1}$ = 44,650	Σx_{c_2} = 1,147 $(\Sigma x_{c_2})^2$ = 1,315,609 $\Sigma x^2_{c_2}$ = 100,959	Σx_{c_3} = 1,478 $(\Sigma x_{c_3})^2$ = 2,184,484 $\Sigma x^2_{c_3}$ = 164,242	Σx_t = 3,343 $(\Sigma x_t)^2$ = 11,175,649 Σx^2_t = 309,851

Step 2: Calculate sum of squares.

Rows:

$$ss_r = \frac{(\Sigma x_{r_1})^2 + (\Sigma x_{r_2})^2}{n_r} - \frac{(\Sigma x_{r_1} + \Sigma x_{r_2})^2}{N}$$

$$= \frac{1{,}560{,}001 + 4{,}384{,}836}{24} - \frac{(1{,}249 + 2{,}094)^2}{48}$$

$$= 247{,}701.54 - 232{,}826.02 = 14{,}875.52$$

Columns:

$$ss_c = \frac{(\Sigma x_{c_1})^2 + (\Sigma x_{c_2})^2 + (\Sigma x_{c_3})^2}{n_c} - \frac{(\Sigma x_{c_1} + \Sigma x_{c_2} + \Sigma x_{c_3})^2}{N}$$

$$= \frac{515{,}524 + 1{,}315{,}609 + 2{,}184{,}484}{16} - \frac{(718 + 1{,}147 + 1{,}478)^2}{48}$$

$$= 250{,}976.06 - 232{,}826.02 = 18{,}150.04$$

Within:

$$ss_w = \Sigma x_t^2 - \frac{\begin{bmatrix}(\Sigma x_{r_1c_1})^2 + (\Sigma x_{r_1c_2})^2 + (\Sigma x_{r_1c_3})^2 + \\ (\Sigma x_{r_2c_1})^2 + (\Sigma x_{r_2c_2})^2 + (\Sigma x_{r_2c_3})^2\end{bmatrix}}{n_w}$$

$$= 309{,}851 \frac{\begin{bmatrix}76{,}729 + 156{,}025 + 332{,}929 + \\ 194{,}481 + 565{,}504 + 811{,}801\end{bmatrix}}{8}$$

$$= 309{,}851 - \frac{2{,}137{,}469}{8} = 42{,}667.37$$

Interaction:

$$ss_i = \frac{\left[\begin{array}{c}(\Sigma x_{r_1c_1})^2 + (\Sigma x_{r_1c_2})^2 + (\Sigma x_{r_1c_3})^2 + \\ (\Sigma x_{r_2c_1})^2 + (\Sigma x_{r_2c_2})^2 + (\Sigma x_{r_2c_3})^2\end{array}\right]}{n_w}$$

$$- \frac{\left[(\Sigma x_{r_1})^2 + (\Sigma x_{r_2})^2\right]}{n_r}$$

$$- \frac{\left[(\Sigma x_{c_1})^2 + (\Sigma x_{c_2})^2 + (\Sigma x_{c_3})^2\right]}{n_c}$$

$$+ \frac{(\Sigma x_t)^2}{N}$$

$$= \frac{\left[\begin{array}{c}76{,}729 + 156{,}025 + 332{,}929 + \\ 194{,}481 + 565{,}504 + 811{,}801\end{array}\right]}{8}$$

$$- \frac{\left[1{,}560{,}001 + 4{,}384{,}836\right]}{24}$$

$$- \frac{\left[515{,}524 + 1{,}315{,}609 + 2{,}184{,}484\right]}{16}$$

$$+ \frac{11{,}175{,}649}{48}$$

$$= [267{,}183.63 - 247{,}701.54 - 250{,}976.06] + 232{,}826.02$$

$$= 1332.05$$

Total: $$ss_t = \Sigma x_t^2 - \frac{(\Sigma x_t)^2}{N}$$

$$= 309{,}851 - \frac{11{,}175{,}649}{48}$$

$$= 77{,}024.98$$

Step 3: Calculate degrees of freedom.

Rows: $R - 1 = 2 - 1 = 1$

Columns: $C - 1 = 3 - 1 = 2$

Within: $N - (R)(C) = 48 - (2)(3) = 42$

Interaction: $(R - 1)(C - 1) = (1)(2) = 2$

Total: $N - 1 = 48 - 1 = 47$

Step 4: Calculate variance estimates.

Rows: $s_r^2 = \frac{ss_r}{df} = \frac{14{,}875.52}{1} = 14{,}875.52$

Columns: $s_c^2 = \frac{ss_c}{df} = \frac{18{,}150.04}{2} = 9{,}075.02$

Within: $s_w^2 = \frac{ss_w}{df} = \frac{42{,}667.37}{42} = 1{,}015.88$

Interaction: $s^2 = \frac{ss_i}{df} = \frac{1{,}332.05}{2} = 666.02$

Step 5: Calculate F ratios.

Rows: $F_r = \frac{s_r^2}{s_w^2} = \frac{14{,}875.52}{1{,}015.88} = 14.64$

Columns: $F_c = \frac{s_c^2}{s_w^2} = \frac{9{,}075.02}{1{,}015.88} = 8.93$

Interaction: $F_i = \frac{s_i^2}{s_w^2} = \frac{666.02}{1{,}015.88} = .65$

Note: Since the within-group variance is an estimate of normal random variation and, therefore, not attributable to the treatment effect, the treatment variance is obtained by dividing the rows, columns, and interaction variances by the within-group, or error, variance.

Accordingly, the critical value of F is obtained in relation to the within-group degrees of freedom.

Step 6: Determine the critical values of F.

Rows: Refer to the F table (Table 8.4) for a .05 level of significance with 1 df for the greater variance estimate and 42 df for the lesser variance estimate (i.e., rows, with a greater variance estimate of 14,875 has 1 df while within, with a lesser variance estimate of 1,015 has 42 df). Thus, the critical value of F is 4.07.

Columns: For 2 df for the greater variance estimate (columns = 9,075) and 42 df for the lesser variance estimate (within = 1,015), the critical value of F is 3.22.

Interaction: For 42 df for the greater variance estimate (within = 1,015) and 2 df for the lesser variance estimate (interaction = 666), the critical value of F is 19.47.

Step 7: Consolidate appropriate information in table form.

SOURCE OF VARIATION	SUM OF SQUARES	DEGREES OF FREEDOM	VARIANCE ESTIMATE	F	CRITICAL F	P
Rows (pressure)	14,875.52	1	14,875.52	14.64	4.07	$<.05$
Columns (speed)	18,150.04	2	9,075.02	8.93	3.22	$<.05$
Within-Cells (error)	42,667.38	42	1,015.88	—	—	—
Interaction	1,332.04	2	666.02	.65	19.47	$>.05$
Total	77,024.98	47				

From the above results, we can conclude that both speed and pressure have a significant effect on print clarity. Accordingly, we can safely conclude that there is not a significant interaction between the two variables (Figure 9.4).

INTERACTION

Significance for Rows, Columns, and Cells

While the critical value of F denotes significance between columns in Example 9.3, it does not specifically indicate where the significance exists. This can be determined by performing individual t tests between columns 1 and 2, 1 and 3, and 2 and 3.

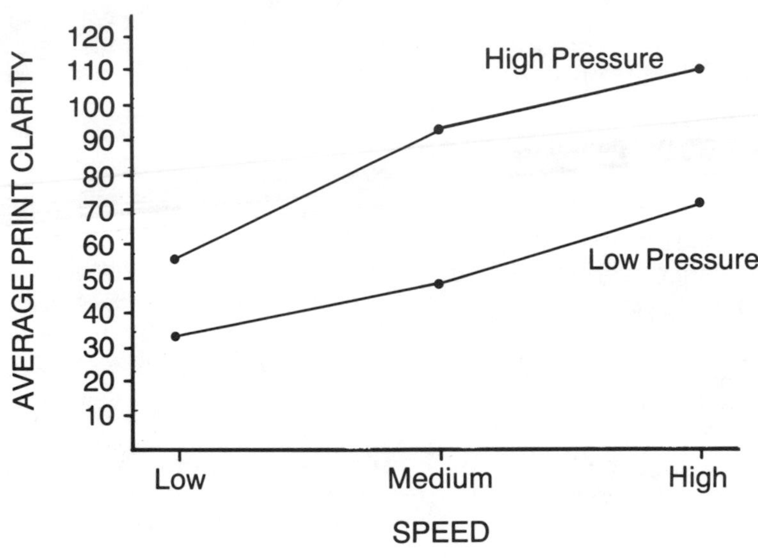

FIGURE 9.4

For significance between columns 1 and 2, we have:

$$t = \frac{\overline{X}_{c_1} - \overline{X}_{c_2}}{\sqrt{\dfrac{S_{1,\,2}^{\;2}}{n_1} + \dfrac{S_{1,\,2}^{\;2}}{n_2}}} = 2.35$$

Additional t tests can also be performed to determine significance between rows 1 and 2, or any two cells.

NONPARAMETRIC DESIGNS

Structure and Effects

Tests of significance usually involve assumptions about the nature, or shape, of the population distributions being tested. The t test and analysis of variance, for example, assume that the population distributions are normal. When this is not the case, or when little is known about the population distribution of the dependent variable, experiments should be designed for nonparametric testing (i.e., non-parameter testing).

The basic structure of a nonparametric factorial experiment is essentially the same as a parametric experiment. For the nonparametric experiment, however, instead of analyzing the variance of a particular sample parameter, the variation of yield effects is studied (i.e., while the manipulation of the independent variable(s) may remain the same, yield becomes the dependent variable which is measured across the various experimental conditions).

For any given number of variables and/or levels, factorial designs, whether parametric or nonparametric, require the same number of runs. For a 3 factor/2 level experiment, therefore, we have 8 runs where each factor, or independent variable, is subjected to 4 runs at the high or + level and 4 runs at the low or – level.

This arrangement is shown graphically as:

	Variable		
Run	A	B	C
1	–		–
2	+	–	–
3	–	+	–
4	+	+	–
5	–	–	+
6	+	–	+
7	–	+	+
8	+	+	+

where the + and – signs symbolize the high and low conditions of the independent variables.

By multiplying the signs across each variable for each run, signs can be assigned for the interaction effect between variables. For example, the sign associated with the AB interaction is positive when both A and B are negative (i.e., – time – is +). Applying these multiplication rules, a table can be constructed for both variables and interactions (Figure 9.5).

As observed in the table of signs (Figure 9.5), mean and divisor information has been included.

	Mean	Run	A	B	C	AB	AC	BC	ABC
	+	1	−	−	−	+	+	+	−
	+	2	+	−	−	−	−	+	+
	+	3	−	+	−	−	+	−	+
	+	4	+	+	−	+	−	−	−
	+	5	−	−	+	+	−	−	+
	+	6	+	−	+	−	+	−	−
	+	7	−	+	+	−	−	+	−
	+	8	+	+	+	+	+	+	+
divisor	8		4	4	4	4	4	4	4

FIGURE 9.5

The mean establishes the baseline for the experimental results and is always positive. This is verified by multiplying the signs across all variables and interactions. The divisor for the mean, therefore, is 8. This is readily discernible when we consider that in order to obtain the average yield for all runs, the sum of the yields would be divided by 8.

The divisor for the individual variables and interactions, however, is 4. This is because each variable and interaction experiences 4 plus and 4 minus conditions for the 8 runs. This results in a net effect of 4 positive and 4 negative conditions divided by the number of conditions (i.e., $[4 + 4] \div 2 = 4$).

Example 9.4
Determine the effect of impact pressure, print speed, and ribbon cartridge configuration on final test print clarity yields.

Step 1: Construct a table of signs to establish the run sequence for the different variable level combinations.

This is usually performed by randomizing the run numbers (i.e., condition 5 as shown in the previous sign table (− − +) could, by selecting random numbers, become the first run condition. Accordingly, previous run condition 1 (− − −) could become the 7th run condition. For the purpose of this example, however, the run sequence in the previous table will be used:

	Mean	Run	P	S	C	PS	PC	SC	PSC	Yield
	+	1	–	–	–	+	+	+	–	60
	+	2	+	–	–	–	–	+	+	72
	+	3	–	+	–	–	+	–	+	54
	+	4	+	+	–	+	–	–	–	68
	+	5	–	–	+	+	–	–	+	52
	+	6	+	–	+	–	+	–	–	83
	+	7	–	+	+	–	–	+	–	45
	+	8	+	+	+	+	+	+	+	80
divisor	8		4	4	4	4	4	4	4	

where P is pressure, S is speed, and C is configuration. In this case, values would be selected for low and high pressure and speed, and cartridge configuration would be classified as type A and B.

Step 2: Subject samples to each of the eight experimental conditions; measuring and recording yield information for each sample group. For this example, the yields listed in the above table will be used to calculate effects and interactions.

Step 3: Calculate main effects for P, S, and C.

The main effects for the independent variables (P, S, and C), as well as the interactions, are computed by summing the yields under each experimental condition and dividing by the divisor. For example, the yield for the first run is 60, but since the P sign for the first run is negative, the yield value is taken as −60. For the second run, since the P sign is positive, the yield value is +72, and so forth.

Pressure Effect:

$$P = \frac{-60 + 72 - 54 + 68 - 52 + 83 - 45 + 80}{4}$$

$$= \frac{92}{4} = [23.0]$$

Speed Effect:

$$S = \frac{-60 - 72 + 54 + 68 - 52 - 83 + 45 + 80}{4}$$

$$= \frac{-20}{4} = [-5]$$

Cartridge Effect:

$$C = \frac{-60 - 72 - 54 - 68 + 52 + 83 + 45 + 80}{4}$$

$$= \frac{6}{4} = [1.5]$$

Step 4: Calculate interaction effects.

Pressure/Speed Interaction:

$$PS = \frac{+60 - 72 - 54 + 68 + 52 - 83 - 45 + 80}{4}$$

$$= \frac{6}{4} = [1.5]$$

Pressure/Cartridge Interaction:

$$PC = \frac{+60 - 72 + 54 - 68 - 52 + 83 - 45 + 80}{4}$$

$$= \frac{40}{4} = [10]$$

Speed/Cartridge Interaction:

$$SC = \frac{+60 + 72 - 54 - 68 - 52 - 83 + 45 + 80}{4}$$

$$= \frac{0}{4} = [0]$$

Pressure/Speed/Cartridge Interaction:

$$PSC = \frac{-60 + 72 + 54 - 68 + 52 - 83 - 45 + 80}{4}$$

$$= \frac{2}{4} = [0.5]$$

Step 5: Arrange effects information in table form.

EFFECTS		
Main Effects	Two Factor Effects	Three Factor Effects
P = 23.0 S = −5.0 C = 1.5	P/S = 1.5 P/C = 10.0 S/C = 0.0	P/S/C = 0.5

Step 6: Combine yields for pressure and cartridge and plot interactions (i.e., sum yields when P is minus and C is minus, when P is plus and C is minus, when P is minus and C is plus, and when P and C are both plus).

	YIELDS			
	P − /C −	P + /C −	P − /C +	P + /C +
	60 54	72 68	52 45	83 80
average	57	70	48.5	81.5

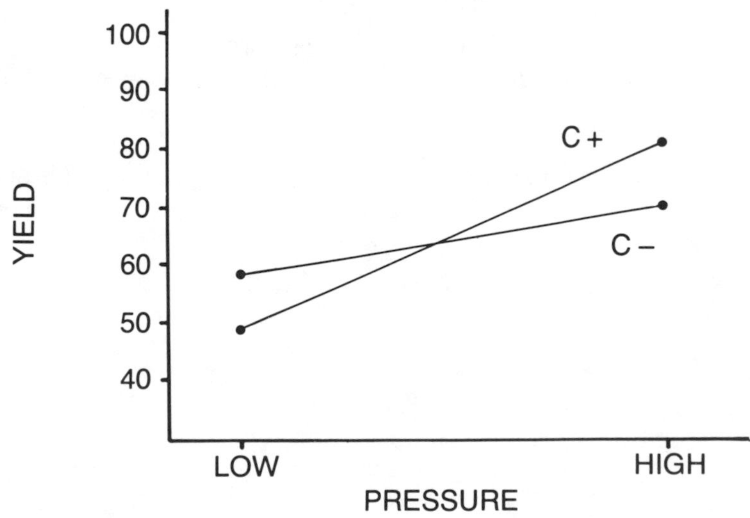

ANALYSIS OF VARIANCE: NONPARAMETRIC

Since a minimum of 1 degree of freedom is needed to compute the variance estimate, nonparametric experimental designs require at least one replicated run (i.e., two runs under each condition or, more conveniently, to simply construct two assumed runs by adding and subtracting one yield point from the results of a single run).

Either approach, however, results in erroneous F values since, for 1 degree of freedom, the variance estimate will equal the sum of squares. For example, for a single replicated run, or two assumed runs, the degrees of freedom for any single effect is equal to 1. Then, since the variance estimate is obtained by dividing the sum of squares by the degrees of freedom, the variance estimate will be equal to the sum of squares. Accordingly, if the error variance is small, as would be expected for a properly designed experiment, the F values could be extremely high; indicating significance where significance may not exist. On the other hand, if additional replicated runs were performed to increase the degrees of freedom and lower the F values, the opposite effect would occur (i.e., failing to indicate significance where significance may exist).

For these reasons, the recommended procedure for nonparametric significance testing is to calculate the sum of squares for each effect (source of variation) and then, based on indicated results, perform t tests to determine if differences between negative and positive yields are significant.

Example 9.5

Using the yield results from the previous example, perform an analysis of variance to determine if the source(s) of variation contributing to yield losses is significant at the .05 level.

Step 1: Calculate the correction factor (c_f) by dividing the sum of the yields squared by the divisor.

$$c_f = \frac{(\Sigma y)^2}{d} = \frac{264{,}196}{8} = 33{,}024$$

Step 2: Calculate the sum of squares (ss) for each effect by summing and squaring the negative and positive yields individually, adding the two results together, dividing by the divisor, and subtracting the correction factor.

$$ss = \frac{(\Sigma - y)^2 + (\Sigma + y)^2}{d} - c_f$$

Pressure:

$$SS_p = \frac{(-211)^2 + (+303)^2}{4} - 33{,}024 = 1{,}058$$

Speed:

$$SS_s = \frac{(-267)^2 + (+247)^2}{4} - 33{,}024 = 50$$

Cartridge:

$$SS_c = \frac{(-254)^2 + (+260)^2}{4} - 33{,}024 = 4.5$$

Pressure/Speed:

$$SS_{ps} = \frac{(-254)^2 + (+260)^2}{4} - 33{,}024 = 4.5$$

Pressure/Cartridge:

$$SS_{pc} = \frac{(-237)^2 + (+277)^2}{4} - 33{,}024 = 200$$

Speed/Cartridge:

$$SS_{sc} = \frac{(-257)^2 + (+257)^2}{4} - 33{,}024 = 0$$

Pressure/Speed/Cartridge:

$$SS_{psc} = \frac{(-256)^2 + (+258)^2}{4} - 33{,}024 = 1$$

Step 3: Compute the total sum of squares by squaring and summing all yield values and subtracting the correction factor.

$$SS_t = \Sigma y^2 - C_f = 34{,}342 - 33{,}024 = 1{,}318$$

Step 4: Perform t tests where sum of squares are sufficiently high (i.e., where variation may be significant).

$$t = \frac{\bar{x}_1 - \bar{x}_2}{\sqrt{\frac{s_{1,2}^2}{n_1} + \frac{s_{1,2}^2}{n_2}}}$$

Where $\bar{x}_1$ and $\bar{x}_2$ are averages for the positive and negative yields.

Pressure:

$$t = \frac{75.75 - 52.75}{4.6} = 5.0$$

Pressure/Cartridge:

$$t = \frac{69.25 - 59.25}{3.2} = 3.1$$

Speed:

$$t = \frac{61.75 - 66.75}{10.2} = 0.4$$

Step 5: List data in table form.

Source of Variation	Sum of Squares	Calculated t Value	Critical t (.05)	P
P (pressure)	1,058	5.0	2.44	$<.05$
S (speed)	50	0.4	2.44	—
C (cartridge)	4	—		
P/S	4	—		
P/C	200	3.1	2.44	$<.05$
S/C	0	—		
P/S/C	1	—		
TOTAL	1,318			

REPLICATED RUNS

Analysis of variance for nonparametric experimental designs requires at least one replicated run (i.e., two runs under each condition). Since this usually imposes hardships in additional resources, however, an alternative procedure is to simply add and subtract one yield point from the obtained yield of each run (i.e., the obtained yield for run 1 in the previous example was 60). Thus, if 1 yield point is added to and subtracted from 60, the result is 59 and 61. These two numbers, then, can be used to represent the required two runs (Figure 9.6).

If two runs are assumed, there are a total of 8 conditions where P, S, and C are either positive or negative. Therefore, when calculating the sum of squares, there will be 8 positive and 8 negative yields. Accordingly, the divisors will double (Figure 9.7).

Notes:

Sum of Squares: The sum of squares and correction factor are calculated in the same manner as shown under Structure and Effects.

Error Sum of Squares: The error sum of squares is obtained by subtracting the sum of the "source of variation" sum of squares from the "total" sum of squares.

Degrees of Freedom: Since the total number of runs is 16, the total degrees of freedom is 15 (i.e., N − 1 = 16 − 1 = 15). Accordingly, since there are two runs for each condition, or source of variation, the degrees of freedom for each source of variation is 1 (i.e., n − 1 = 2 − 1 = 1).

Variance Estimate: The variance estimates are obtained by dividing the sum of squares, for each source of variation, by its associated degrees of freedom. The F value is then obtained by dividing the variance estimates by the error variance.

ACTUAL		ASSUMED RUN/YIELD	
Run	Yield	Run 1	Run 2
1	60	59	61
2	72	71	73
3	54	53	55
4	68	67	69
5	52	51	53
6	83	82	84
7	45	44	46
8	80	79	81

FIGURE 9.6

	I	Run	P	S	C	PS	PC	SC	PSC	Assumed Yield
Assumed Run 1		1	−	−	−	+	+	+	−	59
		2	+	−	−	−	−	+	+	71
		3	−	+	−	−	+	−	+	53
		4	+	+	−	+	−	−	−	67
		5	−	−	+	+	−	−	+	51
		6	+	−	+	−	+	−	−	82
		7	−	+	+	−	−	+	−	44
		8	+	+	+	+	+	+	+	79
Assumed Run 2		1	−	−	−	+	+	+	−	61
		2	+	−	−	−	−	+	+	73
		3	−	+	−	−	+	−	+	55
		4	+	+	−	+	−	−	−	69
		5	−	−	+	+	−	−	+	53
		6	+	−	+	−	+	−	−	84
		7	−	+	+	−	−	+	−	46
		8	+	+	+	+	+	+	+	81
divisor		16		8	8	8	8	8	8	

FIGURE 9.7

Source of Variation	Sum of Squares	Degrees of Freedom	Variance Estimate	F	Critical F(.05)	P
P(pressure)	2,116	1	2,116	1,058	5.32	<.05
S(speed)	100	1	100	50	5.32	<.05
C(cartridge)	9	1	9	4.5	5.32	
P/S	9	1	9	4.5	5.32	
P/C	400	1	400	200	5.32	<.05
S/C	0	1	0	0	—	
P/S/C	1	1	1	.5	239	
Error	16	8	2			
TOTAL	2,651	15				

Exercise Worksheets

Exercise 9.1

ONE-WAY ANALYSIS OF VARIANCE

PROBLEM: Four sample groups are subjected to 4 levels of environmental stress to determine probable causes resulting in post-shipment product failures. Given the following delta performance data, determine, to a .05 level, if a significant difference exists across the specification limits (i.e., a significant source of variation).

SAMPLE GROUPS			
I	II	III	IV
7	5	9	8
4	9	14	2
5	6	18	7
2	11	21	3
6	12	20	10
3	5	12	7
4	8	16	5
1	7	13	6

Exercise 9.1 (continued)

STEP 1: Calculate specified values for each sample group.

	GROUP I	GROUP II	GROUP III	GROUP IV
$n =$	[ ]	[ ]	[ ]	[ ]
$\Sigma x =$	[ ]	[ ]	[ ]	[ ]
$\frac{(\Sigma x)^2}{n} =$	[ ]	[ ]	[ ]	[ ]
$\Sigma x^2 =$	[ ]	[ ]	[ ]	[ ]

STEP 2: Calculate specified values for combined sample groups.

$N =$ [ ]

$\Sigma\Sigma x =$ [ ]

$\Sigma \frac{(\Sigma x)^2}{n} =$ [ ]

$\Sigma\Sigma x^2 =$ [ ]

$\frac{(\Sigma\Sigma x)^2}{N} =$ [ ]

Exercise 9.1 (continued)

STEP 3: Calculate sum of squares.

$$ss_b = \Sigma \frac{(\Sigma x)^2}{n} - \frac{(\Sigma\Sigma x)^2}{N}$$

$$= [\qquad] - [\qquad] = [\qquad]$$

$$ss_w = \Sigma\Sigma x^2 - \Sigma \frac{(\Sigma x)^2}{n}$$

$$= [\qquad] - [\qquad] = [\qquad]$$

$$ss_t = \Sigma\Sigma x^2 - \frac{(\Sigma\Sigma x)^2}{N}$$

$$= [\qquad] - [\qquad] = [\qquad]$$

STEP 4: Calculate degrees of freedom.

ss_b: df $= K\,(\text{groups}) - 1 = [\qquad]$

ss_w: df $= (n_1 - 1) + (n_2 - 1) + (n_3 - 1) + (n_4 - 1)$

$= [\quad] + [\quad] + [\quad] + [\quad] = [\quad]$

ss_t: df $= N - 1 = [\qquad]$

Exercise 9.1 (continued)

STEP 5: Calculate variance estimate.

$$s_b^2 = \frac{ss_b}{df} = \frac{[\quad]}{[\quad]} = [\quad]$$

$$s_w^2 = \frac{ss_w}{df} = \frac{[\quad]}{[\quad]} = [\quad]$$

STEP 6: Calculate F ratio.

$$F = \frac{s_b^2}{s_w^2} = \frac{[\quad]}{[\quad]} = [\quad]$$

STEP 7: Determine critical F value.

df Greater Variance estimate = [ ] Critical F [ ]

df Lesser Variance estimate = [ ]

STEP 8: List specified data in table.

Source of Variation	Sum of Squares	df	Variance Estimate	F	Critical F	Significant Yes	Significant No
Between	[ ]	[ ]	[ ]	[ ]	[ ]	☐	☐
Within	[ ]	[ ]	[ ]				
Total	[ ]	[ ]					

Exercise 9.2

TWO-WAY ANALYSIS OF VARIANCE

PROBLEM: For a robotics part insertion system, investigate the motor torque versus part weight requirements for optimal insertion.

Given the following average and standard deviation values from the 2×3 experimental results listed in Step 1, page 249, plot cell averages to graphically illustrate torque/weight interaction.

			Light Weight	Medium Weight	Heavy Weight
Low	$\bar{x}$	=	.060	0.17	.240
Torque	s	=	.006	.022	.050
High	$\bar{x}$	=	.130	.150	0.17
Torque	s	=	.014	.014	.014

Exercise 9.2 (continued)

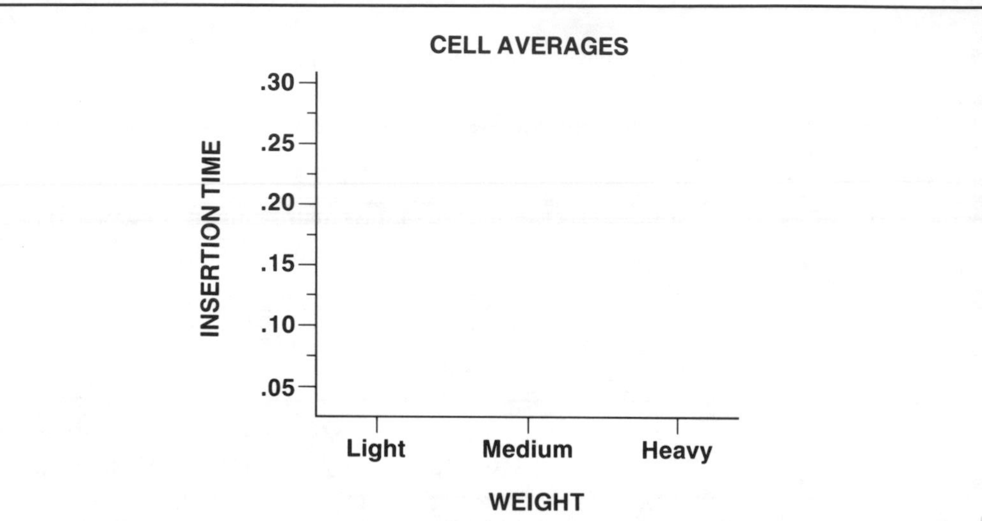

ANALYSIS OF VARIANCE: Given the experimental results in Step 1, perform an analysis of variance and list applicable data in the following variance table (use .05 level of significance).

Source of Variation	Sum of Squares	df	Variance Estimate	Calculated F	Critical F
Rows (torque)	[ ]	[ ]	[ ]	[ ]	[ ]
Columns (weight)	[ ]	[ ]	[ ]	[ ]	[ ]
Interaction	[ ]	[ ]	[ ]	[ ]	[ ]
Within (error)	[ ]	[ ]	[ ]		
Total	[ ]	[ ]			

Exercise 9.2 (continued)

STEP 1: Compute Σx, $(\Sigma x)^2$, and Σx^2 for each cell, column, row, and total.

TORQUE (rows) × **PART WEIGHT** (columns)

	Light	Medium	Heavy	Total
Low	r_1c_1: .058 .062 .065 .056 .068 .051	r_1c_2: .14 .15 .17 .19 .20 .17	r_1c_3: .16 .24 .25 .28 .30 .21	
	$\Sigma x_{r_1c_1}$ = [ ] $(\Sigma x_{r_1c_1})^2$ = [ ] $\Sigma x^2_{r_1c_1}$ = [ ]	$\Sigma x_{r_1c_2}$ = [ ] $(\Sigma x_{r_1c_2})^2$ = [ ] $\Sigma x^2_{r_1c_2}$ = [ ]	$\Sigma x_{r_1c_3}$ = [ ] $(\Sigma x_{r_1c_3})^2$ = [ ] $\Sigma x^2_{r_1c_3}$ = [ ]	Σx_{r_1} = [ ] $(\Sigma x_{r_1})^2$ = [ ] $\Sigma x^2_{r_1}$ = [ ]
High	r_2c_1: .11 .12 .14 .15 .13 .13	r_2c_2: .13 .15 .17 .16 .14 .15	r_2c_3: .15 .16 .17 .18 .19 .17	
	$\Sigma x_{r_2c_1}$ = [ ] $(\Sigma x_{r_2c_1})^2$ = [ ] $\Sigma x^2_{r_2c_1}$ = [ ]	$\Sigma x_{r_2c_2}$ = [ ] $(\Sigma x_{r_2c_2})^2$ = [ ] $\Sigma x^2_{r_2c_2}$ = [ ]	$\Sigma x_{r_2c_3}$ = [ ] $(\Sigma x_{r_2c_3})^2$ = [ ] $\Sigma x^2_{r_2c_3}$ = [ ]	Σx_{r_2} = [ ] $(\Sigma x_{r_2})^2$ = [ ] $\Sigma x^2_{r_2}$ = [ ]
Total	Σx_{c_1} = [ ] $(\Sigma x_{c_1})^2$ = [ ] $\Sigma x^2_{c_1}$ = [ ]	Σx_{c_2} = [ ] $(\Sigma x_{c_2})^2$ = [ ] $\Sigma x^2_{c_2}$ = [ ]	Σx_{c_3} = [ ] $(\Sigma x_{c_3})^2$ = [ ] $\Sigma x^2_{c_3}$ = [ ]	Σx_t = [ ] $(\Sigma x_t)^2$ = [ ] Σx^2_t = [ ]

Exercise 9.2 (continued)

STEP 2: Calculate sum of squares.

Rows:

$$ss_r = \frac{(\Sigma x_{r_1})^2 + (\Sigma x_{r_2})^2}{n_r} - \frac{(\Sigma x_{r_1} + \Sigma x_{r_2})^2}{N}$$

$$= \frac{[\quad\quad] + [\quad\quad]}{[\quad\quad]} - \frac{\left([\quad\quad] + [\quad\quad]\right)^2}{[\quad\quad]}$$

$$= [\quad\quad] - [\quad\quad] = [\quad\quad]$$

Columns:

$$ss_c = \frac{(\Sigma x_{c_1})^2 + (\Sigma x_{c_2})^2 + (\Sigma x_{c_3})^2}{n_c} - \frac{(\Sigma x_{c_1} + \Sigma x_{c_2} + \Sigma x_{c_3})^2}{N}$$

$$= \frac{[\quad\quad] + [\quad\quad] + [\quad\quad]}{[\quad\quad]} - \frac{\left([\quad\quad] + [\quad\quad] + [\quad\quad]\right)^2}{[\quad\quad]}$$

$$= [\quad\quad] - [\quad\quad] = [\quad\quad]$$

Exercise 9.2 (continued)

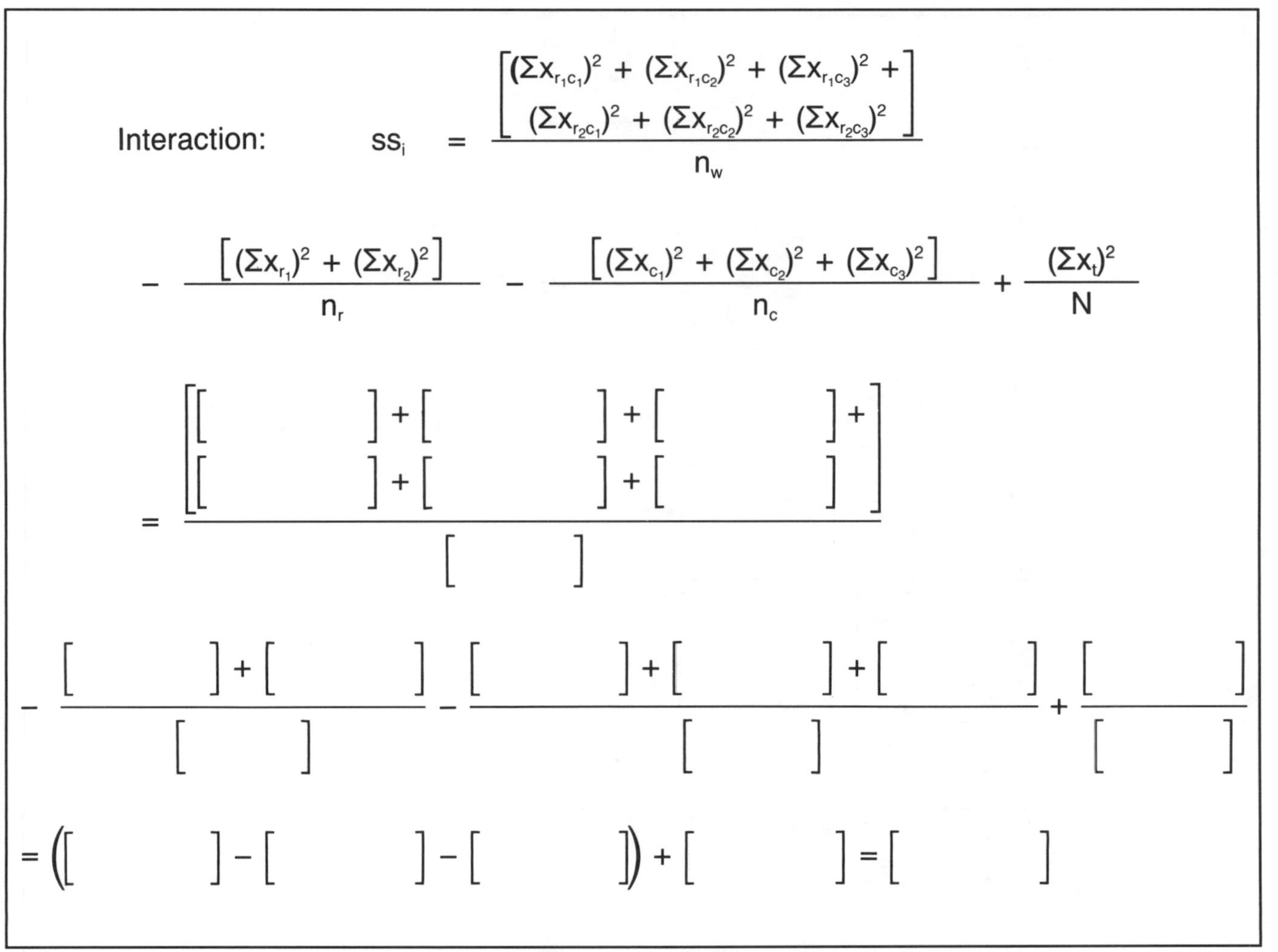

Exercise 9.2 (continued)

Within:

$$ss_w = \Sigma x_t^2 - \frac{\begin{bmatrix}(\Sigma x_{r_1c_1})^2 + (\Sigma x_{r_1c_2})^2 + (\Sigma x_{r_1c_3})^2 + \\ (\Sigma x_{r_2c_1})^2 + (\Sigma x_{r_2c_2})^2 + (\Sigma x_{r_2c_3})^2\end{bmatrix}}{n_w}$$

$$= [\quad] - \frac{\begin{matrix}[\quad] + [\quad] + [\quad] + \\ [\quad] + [\quad] + [\quad]\end{matrix}}{[\quad]}$$

$$= [\quad] - [\quad] = [\quad]$$

Total:

$$ss_t = \Sigma x_t^2 - \frac{(\Sigma x_t)^2}{N}$$

$$= [\quad] - \frac{[\quad]}{[\quad]} = [\quad]$$

Exercise 9.2 (continued)

STEP 3: Calculate degrees of freedom.

Rows : df = R – 1 = [ ]

Columns : df = C – 1 = [ ]

Interaction : df = (R – 1)(C – 1) = [ ]

Within : df = N – (R × C) = [ ]

Total : df = N – 1 = [ ]

STEP 4: Calculate variance estimates.

Rows : $s_r^2 = \frac{ss_r}{df} = \frac{[\quad]}{[\quad]} = [\quad]$

Columns : $s_c^2 = \frac{ss_c}{df} = \frac{[\quad]}{[\quad]} = [\quad]$

Interaction : $s_i^2 = \frac{ss_i}{df} = \frac{[\quad]}{[\quad]} = [\quad]$

Within : $s_w^2 = \frac{ss_w}{df} = \frac{[\quad]}{[\quad]} = [\quad]$

Exercise 9.2 (continued)

STEP 5: Calculate F ratios.

Rows : $F_r = \frac{s_r^2}{s_w^2} = \frac{[\quad]}{[\quad]} = [\quad]$

Columns : $F_c = \frac{s_c^2}{s_w^2} = \frac{[\quad]}{[\quad]} = [\quad]$

Interaction : $F_i = \frac{s_i^2}{s_w^2} = \frac{[\quad]}{[\quad]} = [\quad]$

STEP 6: Determine critical values of F.

Rows : df greater variance estimate = [ ]

df lesser variance estimate = [ ]

F = [ ]

Columns : df greater variance estimate = [ ]

df lesser variance estimate = [ ]

F = [ ]

Interaction : df greater variance estimate = [ ]

df lesser variance estimate = [ ]

F = [ ]

Exercise 9.3

NONPARAMETRIC ANALYSIS OF VARIANCE

PROBLEM: The major parameters affecting product performance are input power (P), signal strength (S), and coupling capacitance (C). Given the following experimental data, determine: 1) which of the three parameters have the greatest effect on yield loss; 2) if there are interactions affecting product performance; and 3) if observed yield differences are attributable to expected random variation or significant at a .05 level.

	Mean	Run	P	S	C	PS	PC	SC	PSC	Yield
	+	4	–	–	–	+	+	+	–	74
	+	8	+	–	–	–	–	+	+	76
	+	2	–	+	–	–	+	–	+	83
	+	1	+	+	–	+	–	–	–	74
	+	3	–	–	+	+	–	–	+	68
	+	5	+	–	+	–	+	–	–	84
	+	7	–	+	+	–	–	+	–	76
	+	6	+	+	+	+	+	+	+	88
divisor	8		4	4	4	4	4	4	4	

Exercise 9.3 (continued)

STEP 1: Compute the main effects of P, S, and C by summing their respective positive and negative yields and dividing by the divisor.

Main effect of P:

$$\frac{[\quad]+[\quad]+[\quad]+[\quad]+[\quad]+[\quad]+[\quad]+[\quad]}{[\quad]} = \frac{[\quad]}{[\quad]} = [\quad]$$

Main effect of S:

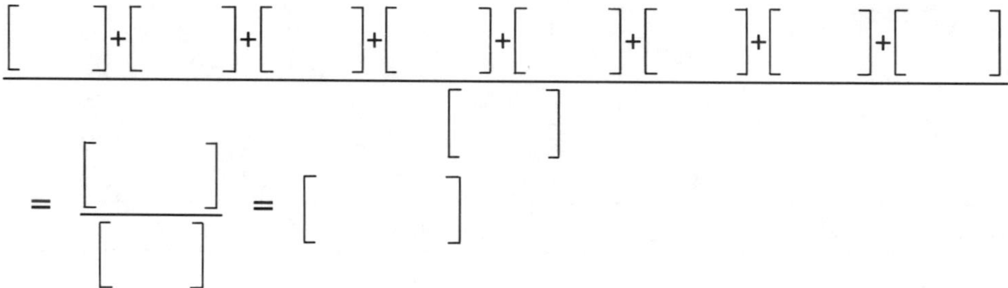

Main effect of C:

$$\frac{[\quad]+[\quad]+[\quad]+[\quad]+[\quad]+[\quad]+[\quad]+[\quad]}{[\quad]} = \frac{[\quad]}{[\quad]} = [\quad]$$

Exercise 9.3 (continued)

STEP 2: Compute the interaction effects of P/S, P/C, S/C, and P/S/C in the same manner as the main effects.

Interaction effect of P/S

$$\frac{[\quad]+[\quad]+[\quad]+[\quad]+[\quad]+[\quad]+[\quad]+[\quad]}{[\quad]} = \frac{[\quad]}{[\quad]} = [\quad]$$

Interaction effect of P/C

$$\frac{[\quad]+[\quad]+[\quad]+[\quad]+[\quad]+[\quad]+[\quad]+[\quad]}{[\quad]} = \frac{[\quad]}{[\quad]} = [\quad]$$

Interaction effect of S/C

$$\frac{[\quad]+[\quad]+[\quad]+[\quad]+[\quad]+[\quad]+[\quad]+[\quad]}{[\quad]} = \frac{[\quad]}{[\quad]} = [\quad]$$

Exercise 9.3 (continued)

Interaction effect of P/S/C:

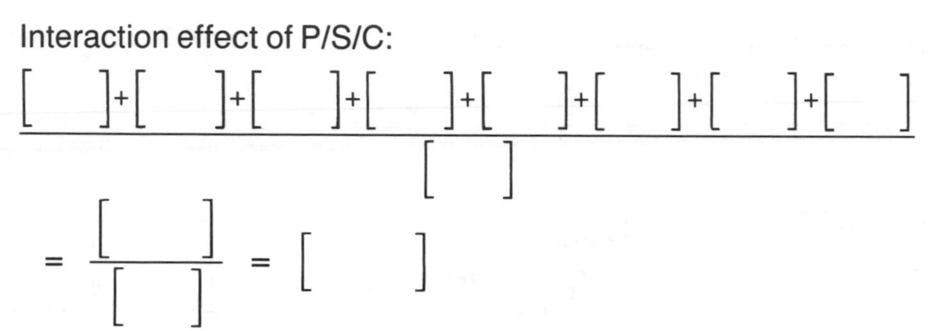

STEP 3: Complete a table of effects.

MAIN EFFECTS		TWO FACTOR EFFECTS		THREE FACTOR EFFECT	
P	= [ ]	P/S	= [ ]	P/S/C	= [ ]
S	= [ ]	P/C	= [ ]		
C	= [ ]	S/C	= [ ]		

STEP 4: Compute average yield for the greater interaction effect.

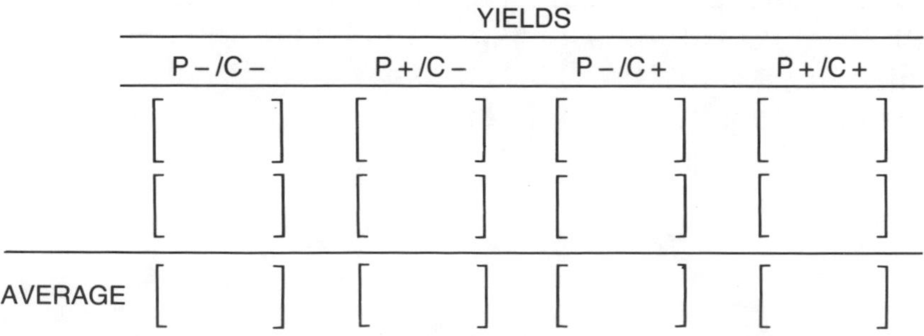

	YIELDS			
	P – /C –	P + /C –	P – /C +	P + /C +
	[ ]	[ ]	[ ]	[ ]
	[ ]	[ ]	[ ]	[ ]
AVERAGE	[ ]	[ ]	[ ]	[ ]

Exercise 9.3 (continued)

STEP 5: Plot interaction effect of P/C.

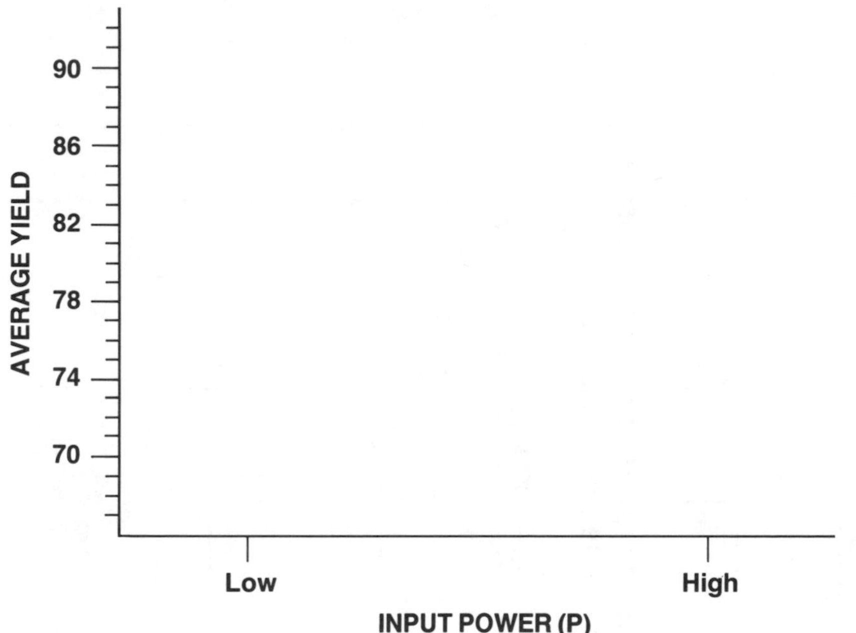

ANALYSIS OF VARIANCE

Perform analysis of variance to determine if observed yield differences for the various experimental conditions are significant (at a .05 level) and to determine the major source of variation resulting in yield loss.

Exercise 9.3 (continued)

STEP 1: Compute the correction factor.

$$c_f = \frac{(\Sigma y)^2}{d} = \frac{[\quad]}{[\quad]} = [\quad]$$

STEP 2: Compute the sum of squares for main and interaction effects and list in table under Step 5.

$$\text{Sum of squares (ss)} = \frac{(\Sigma - y)^2 + (\Sigma + y)^2}{d} - c_f$$

$$ss_p = \frac{[\quad] + [\quad]}{[\quad]} - [\quad] = [\quad]$$

$$ss_s = \frac{[\quad] + [\quad]}{[\quad]} - [\quad] = [\quad]$$

$$ss_c = \frac{[\quad] + [\quad]}{[\quad]} - [\quad] = [\quad]$$

Exercise 9.3 (continued)

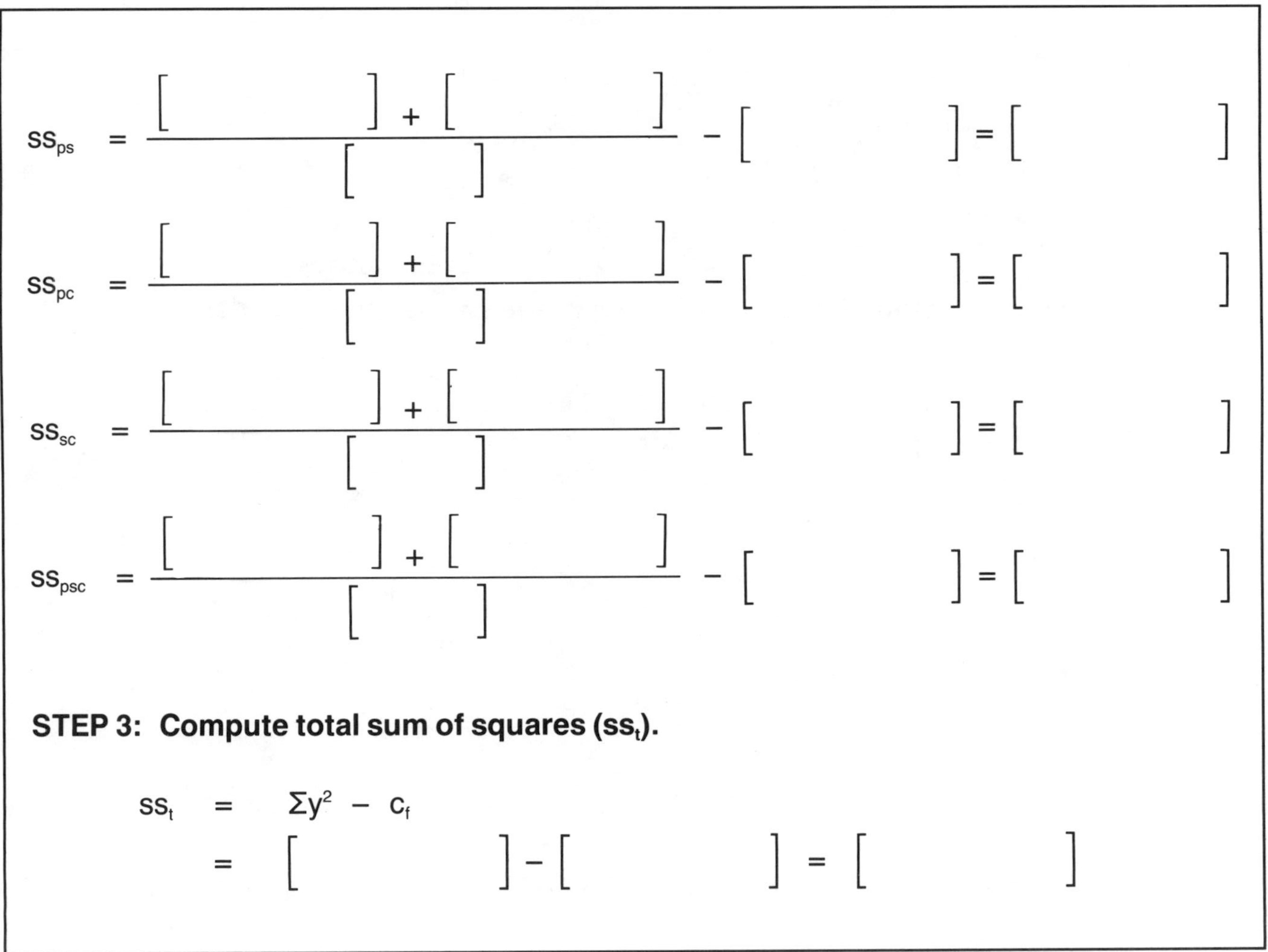

Exercise 9.3 (continued)

STEP 4: Compute t values for effects which indicate probable significance.

Sub-Step 1: Calculate required values for pooled variance estimates.

	P/C EFFECT		P EFFECT	
	+ Yields	– Yields	+ Yields	– Yields
$\bar{x}$ =	[ ]	[ ]	[ ]	[ ]
Σx =	[ ]	[ ]	[ ]	[ ]
$(\Sigma x)^2$ =	[ ]	[ ]	[ ]	[ ]
Σx^2 =	[ ]	[ ]	[ ]	[ ]

Sub-Step 2: Calculate t values using following t and pooled variance formulas.

$$t = \frac{\bar{x}_1 - \bar{x}_2}{\sqrt{\frac{s^2_{1,2}}{n_1} + \frac{s^2_{1,2}}{n_2}}}$$

Where

$$s^2_{1,2} = \frac{\left[\Sigma x_1^2 - \frac{(\Sigma x_1)^2}{n_1}\right] + \left[\Sigma x_2^2 - \frac{(\Sigma x_2)^2}{n_2}\right]}{n_1 + n_2 - 2}$$

Exercise 9.3 (continued)

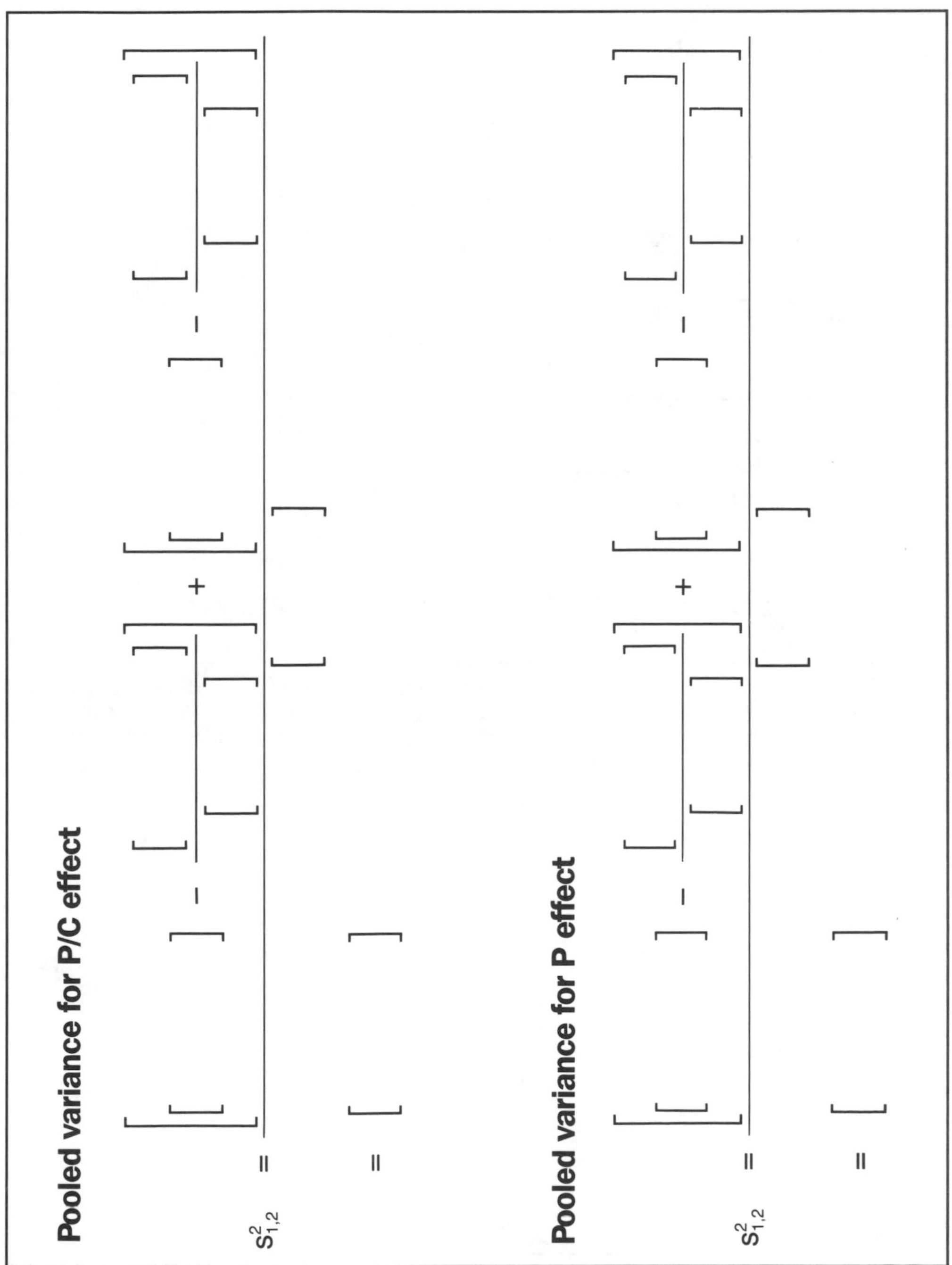

Exercise 9.3 (continued)

t value for P/C effect.

$$t = \frac{[\qquad] - [\qquad]}{\sqrt{\dfrac{[\qquad]}{[\qquad]} + \dfrac{[\qquad]}{[\qquad]}}} = [\qquad]$$

t value for P effect.

$$t = \frac{[\qquad] - [\qquad]}{\sqrt{\dfrac{[\qquad]}{[\qquad]} + \dfrac{[\qquad]}{[\qquad]}}} = [\qquad]$$

STEP 5: List data and critical t values in table form.

Source of Variation	Sum of Squares	Calculated t Value	Critical t Value	Significant Yes	Significant No
P	[ ]	[ ]	[ ]	☐	☐
S	[ ]				
C	[ ]				
P/S	[ ]				
P/C	[ ]	[ ]	[ ]	☐	☐
S/C	[ ]				
P/S/C	[ ]				

BIBLIOGRAPHY

Anderson, V. L., and R. A. McLean. *Design of Experiments — A Realistic Approach.* New York: Marcel Dekker, 1974.

Box, G. E. P., W. G. Hunter, and J. S. Hunter. *Statistics for Experimenters — An Introduction to Design, Data Analysis, and Model Building.* New York: John Wiley & Sons, 1978.

Calabro, S. R. *Reliability Principles and Practices.* New York: McGraw-Hill Book Co., 1962.

Ferguson, G. A. *Statistical Analysis in Psychology and Education,* 5th ed. New York: McGraw-Hill Book Co., 1981.

Grant, E. L. *Statistical Quality Control,* 3rd ed. New York: McGraw-Hill Book Co., 1964.

Juran, J. M., and F. M. Gryna, Jr. *Quality Planning and Analysis.* New York: McGraw-Hill Book Co., 1970.

Kurtz, A. K., and S. T. Mayo. *Statistical Methods in Education and Psychology.* New York: Springer – Verlag, 1979.

Matheson, D. W., R. L. Bruce, and K. L. Beauchamp. *Introduction to Experimental Psychology,* 2nd ed. New York: Holt, Rinehart, and Winston, 1974.

Odeh, R. E., and M. Fox. *Sample Size Choice — Charts for Experiments with Linear Models.* New York: Marcel Dekker, 1975.

Rowntree, D. *Statistics Without Tears — A Primer for Non-Mathematicians.* New York: Charles Scribner's Sons, 1981.

Note: Worked answer sheets for all exercises may be obtained by contacting *Publishing Assistant, ASQC Quality Press, 310 West Wisconsin Avenue, Milwaukee, Wisconsin 53203, 414/272-8575.*